T0317584

.

Hydrogen Generation, Storage, and Utilization

Hydrogen Generation, Storage, and Utilization

Jin Zhong Zhang

Jinghong Li

Yat Li

Yiping Zhao

Co-Published by John Wiley & Sons, Inc., Hoboken, New Jersey; and ScienceWise Publishing. Published simultaneously in Canada.

For general information on our other products and services or for technical support, please contact our Customer Care Department within the United States at (800) 762-2974, outside the United States at (317) 572-3993 or fax (317) 572-4002.

Wiley also publishes its books in a variety of electronic formats. Some content that appears in print may not be available in electronic formats. For more information about Wiley products, visit our web site at www.wiley.com.

Library of Congress Cataloging-in-Publication Data:

Zhang, Jin Z.
   Hydrogen generation, storage, and utilization / Jin Zhong Zhang, Jinghong Li, Yat Li, Yiping Zhao.
     pages cm
   Includes bibliographical references and index.
   ISBN 978-1-118-14063-5 (cloth)
  1. Hydrogen–Storage.   2. Energy storage.   3. Hydrogen as fuel. I. Li, Yat. II. Li, Jinghong.
III. Zhao, Yiping. IV. Title.
   TP245.H9Z43   2014
   665.8'1–dc23
                            2013046059

Printed in the United States of America

ISBN: 9781118140635

10 9 8 7 6 5 4 3 2 1

Contents

Preface

Hydrogen is the most abundant element in the universe and accounts for about 75% of the known mass of the universe. Both the hydrogen element (or atom, H) and hydrogen molecule (H_2) have many unique chemical and physical properties. Atomic hydrogen is the smallest and lightest element. Hydrogen is not only an important chemical in itself, but also a major component of many important molecules, such as water, hydrocarbons, polymers, proteins, and deoxyribonucleic acid (DNA). For most organic and biological compounds, hydrogen is the major component in terms of number of atoms per molecule, and it only forms a relatively strong single bond with atoms such as carbon (C), nitrogen (N), and oxygen (O).

Hydrogen is also a major element in many known stars and planets. Stars spend about 90% of their lifetime fusing hydrogen to produce helium in high temperature and high pressure reactions near the core. Thus, hydrogen is a critical element for the very existence of the universe on the grander scale. In particular, hydrogen is a critical energy source, since the nuclear fusion of hydrogen in the sun provides all the energy and supports all the activities on earth.

When two hydrogen atoms combine, they form a stable molecule, H_2, with a single and strong chemical bond. The bond dissociation energy is 436 kJ/mole or 4.52 eV. The equilibrium bond length is 0.074 nm or 0.74 Å. H_2 is a stable molecule but can react with a number of elements and molecules under appropriate conditions. One example is reaction with oxygen (O_2) to form water. Such a reaction can release 241.826 kJ/mol energy, and is considered to be an environmentally clean energy release process.

Among the many interesting properties of H_2, its potential use as a clean and renewable fuel has attracted significant attention recently, especially given the rapidly increasing demand for energy and deteriorating environmental issues associated with use of fossil fuel on a global scale. H_2 is arguably an ideal fuel in many ways. First, when H_2 is consumed for energy production, the only byproduct is water, which is clean and useful. Second is the abundance of hydrogen and its potential low cost. The potential of hydrogen as a clean energy carrier is the primary focus of this book.

Although hydrogen is abundant in Earth, in majority, it exists in stable and strong chemical bonds with atoms such as C, N, and O. To extract hydrogen from those compounds not only requires external energy input, but also a specific chemical reaction route. How to make such reactions simple, accessible, environmentally sustainable, and energetically efficient represents one of the greatest challenges. In addition, although hydrogen has the highest mass energy density among all the chemical fuels, it has almost the lowest volumetric energy density. Thus, another giant challenge for hydrogen fuel is how to store hydrogen with improved volumetric energy density while keeping the mass energy density high. Moreover, hydrogen poses several safety concerns, from potential detonations and fires in air, to being an asphyxiant in its pure, oxygen-free form. Thus, hydrogen must be handled with extreme care and caution. This also raises concerns about its safe use in energy applications.

In this book, we cover some of the fundamental properties of hydrogen and recent developments in the generation, storage, and utilization of hydrogen as a potential sustainable energy carrier. Chapter 1 provides a brief overview of the basic properties of the hydrogen element and hydrogen molecule, including their electronic structure, optical and magnetic properties, and related safety precautions. Chapters 2–4 cover hydrogen generation. Chapter 2 focuses on hydrocarbons as a source for hydrogen generation through techniques such as steam reforming. Chapter 3 concentrates on solar hydrogen generation with emphasis on water splitting using photocatalytic and photoelectrochemical methods. Chapter 4 highlights other methods for hydrogen generation with focus on biological approaches. Chapters 5–7 cover hydrogen storage, with Chapter 5 focusing on established methods based on compression and cryogenics, Chapter 6 on chemical storage based on metal hydrides and hydrocarbon, and Chapter 7 on physical storage using nanostructured and porous materials. Chapters 8–10 discuss several aspects of hydrogen utilization, including combustion (Chapter 8), fuel cells (Chapter 9), and chemical processes (Chapter 10). For each topic, we provide historic background and highlight significant and/or recent developments. Our objective is not to be comprehensive, as the literature is huge and fast growing, but rather to be illustrative using examples.

While there are many challenges at present for using hydrogen as an energy carrier, encouraging progress has been steadily made. It is highly possible that with further research and development, one day hydrogen will become a major part of the energy enterprise.

JIN ZHONG ZHANG
JINGHONG LI
YAT LI
YIPING ZHAO

Acknowledgments

We would like to thank our mentors and many colleagues, collaborators, postdoctors, and students who have contributed directly or indirectly to the writing of this book through discussion, collaboration, and research work. An incomplete list of people to whom we wish to express our gratitude include: Guozhong Cao, Jason Cooper, Elder De La Rosa, Bob Fitzmorris, Jinghua Guo, Min Guo, Charles B. Harris, Greg Hartland, Yuping He, Eric J. Heller, Jennifer Hensel, Jianhua Hu, Dan Imre, Dongling Ma, George Larsen, Shuit-Tong Lee, Steve Leone, Can Li, Yadong Li, Tim Lian, Yichuan Ling, Gang-yu Liu, Jun Liu, Liping Liu, Tianyu Liu, Toh-Ming Lu, Tzarara Lopez-Luke, Sam Mao, Matthew D. McCluskey, Umapada Pal, Fang Qian, Wilson Smith, Xinman Tu, Lionel Vayssieres, Changchun Wang, Dunwei Wang, Gongming Wang, Gwp-Ching Wang, Hanyu Wang, Hua Wang, Heli Wang, Abe Wolcott, Damon Wheeler, Stan Wong, Kui Yu, and Zhongping Zhang.

We wish to thank several funding agencies for partial financial support to research in our labs over the years and our time spent on writing this book, including the U.S. National Science Foundation, the U.S. Department of Energy, and the National Natural Science Foundation, and the Ministry of Science and Technology of China.

We are grateful to our families for their support, love, and understanding.

We wish to thank the book editor, Mr. Dennis Couwenberg, for his initiative, patience, and support for this book project.

1

Introduction to Basic Properties of Hydrogen

1.1 BASICS ABOUT THE HYDROGEN ELEMENT

Hydrogen is known as the most abundant element in the universe. It accounts for about 75% of the known mass of the universe. Hydrogen is a major element in many known stars and planets. For example, stars, when formed in the present Milky Way galaxy, are composed of about 71% hydrogen and 27% helium, as measured by mass, with a small fraction of heavier elements [1]. Stars spend about 90% of their lifetime fusing hydrogen to produce helium in high temperature and high pressure reactions near the core. Thus, hydrogen is a critical element for the very existence of the universe.

Both the hydrogen atom (H) and hydrogen molecule (H_2) have many unique chemical and physical properties. Hydrogen is also a major component of many important molecules, such as water, hydrocarbons, proteins, and DNA. It is safe to say that there would be no life if there were no hydrogen.

The atomic hydrogen is the smallest and lightest element. The hydrogen atom consists of one proton (H^+) and one electron, with no neutrons, and is usually denoted as ^1H or just H (also named as protium sometimes). Hydrogen has two common isotopes, deuterium (D or ^2H) and tritium (T or ^3H), that contain one and two neutrons, respectively, in addition to the one proton

Hydrogen Generation, Storage, and Utilization, First Edition. Jin Zhong Zhang, Jinghong Li, Yat Li, and Yiping Zhao.
© 2014 John Wiley & Sons, Inc. Published 2014 by John Wiley & Sons, Inc.

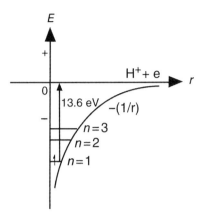

FIGURE 1.1   Relevant energy level of the ground electronic state of the H atom and its ionized state ($H^+ + e$). E is energy, n is the principal quantum number, r is the distance between the electron and proton; $-(1/r)$ is the Coulombic attraction between the electron and proton; and 13.6 eV corresponds to the ionization energy of the H atom from its ground electronic state ($n = 1$ or 1s atomic orbital).

and electron that H contains. The abundance is 99.895%, 0.015%, and trace amount, respectively, for H, D, and T. While the mass differs significantly among the three isotopes, their electronic structures and properties are very similar since the neutrons have essentially no effect on the electronic properties that are mainly determined by the electron and proton. Other highly unstable nuclei (^4H to ^7H) have been synthesized in the laboratory but not observed in nature.

The ionization energy for H atom is 13.6 eV or 1312.0 kJ mol^{-1}, equivalent to a photon energy of 92 nm. Thus, H atom is highly stable under normal conditions. The ionized form of the H atom is the proton, H^+, which has many interesting and unique properties of its own. It is the lightest and smallest atomic ion. Figure 1.1 shows the relevant energy levels for the ground electronic state of H atom relative to its ionized state ($H^+ + e$). In water, the proton is in the form of H_3O^+ and plays a critical role in many biological processes. The proton is also related to acids and bases, which are two essential classes of compounds in chemistry and important for chemical industry.

Hydrogen atoms are reactive and can be combined with many elements to form a huge number of different compounds, including most organic and biological compounds, such as hydrocarbons, polymers, proteins, and DNA. For most organic compounds, the hydrogen is bound to the atoms of carbon and, to a lesser degree, nitrogen, oxygen, or other atoms, such as phosphorus and sulfur. The H atom only forms a relatively strong single bond with these atoms.

1.2   BASICS ABOUT THE HYDROGEN MOLECULE

When two hydrogen atoms combine, they form a stable molecule, H_2, with a single and strong covalent bond. The equilibrium bond length is 0.74 Å. The bond dissociation energy is 4.52 eV or 436 kJ mol^{-1}.

Extensive experimental and theoretical studies have been done on H_2 in terms of its electronic structures, optical properties, magnetic properties, and reactivity with other elements or compounds. Its small size and light mass make it convenient for theoretical and computational studies. For example, potential energy surfaces (PES) or curves for many electronic states of H_2 have been calculated with high accuracy [2, 3]. Figure 1.2 shows some examples of PES of low-lying electronic states of H_2 [4, 5]. The ground electronic state and the first few excited states are all bound with respect to the bond distance between the two hydrogen atoms.

Because the large energy difference between the ground and first excited electronic states of H_2 (near 12 eV), there is no absorption of visible or UV light by H_2, thus H_2 gas is colorless. H_2 does absorb light in the vacuum UV (VUV) region of the spectrum. Since the three lowest excited electronic states are all bound, they are expected to be relatively long-lived and lead to fluorescence when excited by light in the VUV region.

Molecular hydrogen has interesting magnetic properties, mainly due to its nuclear spin properties. There are two different spin isomers of H_2, ortho and

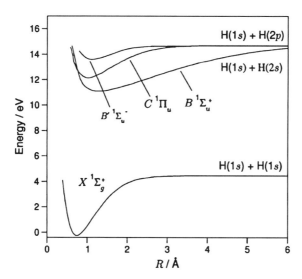

FIGURE 1.2   Examples of several low-lying PES of H_2. *Source*: Reproduced with permission from Flemming et al. [4].

para, which differ by the relative spin of their nuclei. In the orthohydrogen form, the spins of the two protons are parallel to each other and form a triplet state with a molecular spin quantum number of 1 ([1/2] + [1/2]). In the parahydrogen form, the proton spins are antiparallel to each other and form a singlet state with a molecular spin quantum number of 0 ([1/2] − [1/2]). At standard temperature and pressure, H_2 gas contains about 75% of the ortho form and 25% of the para form, known as the normal form. The ortho form has a higher energy than the para form, and is thus unstable and cannot be purified. The ortho/para ratio depends on temperature, and decreases with decreasing temperature. This ratio in condensed H_2 is an important consideration in the preparation and storage of liquid hydrogen (see Chapter 5), since the conversion from ortho to para is exothermic and produces enough heat to evaporate some of the hydrogen liquid, leading to loss of liquefied material. The interconversion between the two forms and hydrogen cooling are often facilitated by catalysts such as ferric oxide, activated carbon, or some nickel compounds.

H_2 is a stable molecule but can react with a number of elements and molecules under certain conditions. One classic example is reaction with oxygen (O_2) to form water. Such a reaction can be carried out by way of combustion, which is fast and violent, as will be discussed in more detail in Chapter 8.

Among the many interesting properties of H_2, its potential use as a clean and renewable fuel has attractive considerable attention, especially given the increasing demand of energy and adverse environmental impact associated with use of fossil fuel. H_2 is an ideal fuel in several ways, including clean byproduct, water, in energy production process and abundance, making it potentially low cost. For example, Chapter 3 covers recent research efforts on photoelectrochemical water splitting for hydrogen generation. However, there are currently some major obstacles toward the practical use of hydrogen as a fuel, including hydrogen generation, storage, transport, and utilization. Efficiency and cost are two important factors to consider for each of these aspects. Safety is another factor of concern.

A molecular ion, H_3^+, has been found in interstellar medium (ISM), which is generated by ionization of H_2 from cosmic rays. It is one of the most stable ions in the universe. The neutral form, H_3, is not stable and can only exist in an excited state for a short period of time.

1.3 OTHER FUNDAMENTAL ASPECTS OF HYDROGEN

Besides forming the hydrogen molecule, H_2, the hydrogen element is involved in the formation of a large number of compounds, including water, hydrocarbons, and many important biological molecules, such as proteins and

DNAs. Water is essential for life, and the water molecule contains two hydrogen atoms and one oxygen atom, bound together in a bent geometry. The H–O–H bond angle is 104.5° in the ground electronic state. The bent geometry results in a net permanent dipole moment of the water molecule. Hydrogen bonding between water molecules is another unique feature of water. The combination of hydrogen bonding and permanent dipole moment are largely responsible for the fact that water is a liquid at room temperature and 1 atm, a property critical for life. If we count the two lone pairs of electrons, water molecule has a near tetrahedral geometry with the oxygen atom in the center. In its ground electronic state, the water molecule has three vibrational frequencies: symmetric stretch (3657 cm^{-1}), antisymmetric stretch (3756 cm^{-1}), and bending (1595 cm^{-1}), based on gas phase data. The relatively high vibrational frequencies, in conjunction with fast rotation, are important for its role as a common solvent and other significant properties, such as heat conduction.

Most organic molecules contain hydrogen atoms. Examples include saturated and unsaturated hydrocarbons, aromatic compounds, lipids, alcohols, ethers, and esters. Many small drug molecules, polymers, and petrochemicals, such as gasoline, are examples of important molecules that contain hydrogen.

Almost all biological molecules contain hydrogen, for example, proteins and DNAs. Of course, complex structures, such as cells, viruses, bacteria, and tissues, all contain hydrogen as an essential element in their constituent components.

1.4 SAFETY AND PRECAUTIONS ABOUT HYDROGEN

Hydrogen poses several safety concerns, from potential detonation and fire in air to being an asphyxiant in its pure, oxygen-free form. For instance, as a cryogen, liquid hydrogen presents dangers associated with very cold liquids. Hydrogen "dissolved" in metals can lead to cracks and explosions. Hydrogen gas in air may spontaneously ignite, and the detonation parameters, for example, critical detonation pressure and temperature, strongly depend on the container geometry. Thus, hydrogen must be handled with extreme care and caution in gaseous or liquid form.

REFERENCES

1. Irwin, J.A. *Astrophysics: Decoding the Cosmos*, John Wiley & Sons, Chichester, 2007.

2. Kolos, W., Rychlewski, J. Unusual double-minimum potential-energy curves: The H and G$^3\Sigma_g^+$ states of the hydrogen molecule. *Journal of Molecular Spectroscopy* **1990**, *143*(2), 212–230.

3. Kolos, W., Rychlewski, J. Adibatic potential-energy curves for the B and $E^3S_u^+$ states of the hydrogen molecule. *Journal of Molecular Spectroscopy* **1990**, *143*(2), 237–250.

4. Flemming, E., Wilhelmi, O., Schmoranzer, H., Glassmaujean, M. Coherence effects in the polarization of Lyman-Alpha fluorescence following photodissociation of H_2 and D_2. *Journal of Chemical Physics* **1995**, *103*(10), 4090–4096.

5. Clark, A.P., Brouard, M., Quadrini, F., Vallance, C. Atomic polarization in the photodissociation of diatomic molecules. *Physical Chemistry Chemical Physics* **2006**, *8*(48), 5591–5610.

2

Hydrocarbons for Hydrogen Generation

2.1 BASICS ABOUT HYDROCARBONS

Hydrocarbons generally refer to a class of organic compounds that contain hydrogen and carbon atoms. Saturated hydrocarbons only contain C–H and C–C single bonds, while unsaturated hydrocarbons contain C=C double or C≡C triple bonds besides single bonds. Hydrocarbons are very important for many applications, especially in the petroleum industry. The hydrogen content in hydrocarbons is often high, when measured in the form of atomic percentage. Methane (CH_4), the major component of natural gas, has a hydrogen mass content of 25%, while propane ($CH_3CH_2CH_3$) contains about 18% of hydrogen by mass. The hydrogen content usually decreases with increasing hydrocarbon molecular weight or size.

Under ambient conditions (one atmosphere pressure and room temperature), hydrocarbons can exist as gas, liquid, or solid, depending on molecular weight or size. Usually, lighter or smaller molecules are found in gas phase, while heavier and larger ones are in liquid or solid phase. This is due to strong intermolecular interaction in large molecules since they are more polarizable or have larger induced dipole moments.

About 95% of hydrogen produced today comes from carbonaceous raw materials, primarily fossil in origin. Hydrogen is generated from hydrocarbons by first breaking the C–H bond. Because this is a relatively strong

Hydrogen Generation, Storage, and Utilization, First Edition. Jin Zhong Zhang, Jinghong Li, Yat Li, and Yiping Zhao.
© 2014 John Wiley & Sons, Inc. Published 2014 by John Wiley & Sons, Inc.

bond, significant energy is required to break it, which is often supplied by heating to high temperature. Like in many chemical reactions, catalysis can be used to weaken the C–H bond, facilitating the bond breaking for hydrogen generation. One of the most common approaches is steam reforming, which involves hydrocarbons and water steam for hydrogen generation, as discussed next.

2.2 STEAM METHANE REFORMING

Steam reforming of natural gas (CH_4) is the most common method for making commercial bulk hydrogen. Most of the hydrogen manufactured in the United States today is from steam reforming of natural gas. At high temperatures (1000–1400 K or 700–1100°C), steam (water vapor) reacts with methane to generate carbon monoxide (CO) and H_2 based on the following reaction:

$$CH_4 + H_2O \rightarrow CO + 3 H_2. \tag{2.1}$$

The standard heat of reaction for this reaction is $\Delta H_{298} = +206.1$ kJ/mol and it is endothermic, requiring external energy input. Even though the reaction is favored at low pressure based on Le Châtelier's principle, it is usually conducted at high pressure (20 atm) since high pressure H_2 is the most marketable product, and purification based on pressure swing adsorption (PSA) works better at higher pressures (e.g., 700–850°C) [1]. The external heat needed is supplied by combustion of a fraction of the incoming natural gas or from burning waste gases. The energy conversion efficiency, defined as the ratio of hydrogen out/energy input, for large-scale reformers is typically in the 75–80% range. Methane and water steam react in tubes filled with catalysts, and the steam-to-methane mass ratio is 3 or higher to avoid carbon buildup or "coking" on the catalyst.

Factors that affect the steam reforming reaction include pressure, temperature, and catalyst used. Usually Ni or the noble metals Ru, Rh, Pd, Ir, and Pt are used as the active catalysts. Due to its low cost, Ni is the most widely used catalyst even though it is less active and usually more prone to deactivation by, for example, carbon formation or oxidation [1]. The activity of a catalyst usually increases with the metal surface area, and thus the catalytic activity benefits from a high dispersion of metal particles. The metal catalysts are usually supported on oxides such as MgO or Al_2O_3, and the support can have substantial influence on the dispersion and catalytic activities of the catalysts and thereby the reaction performance.

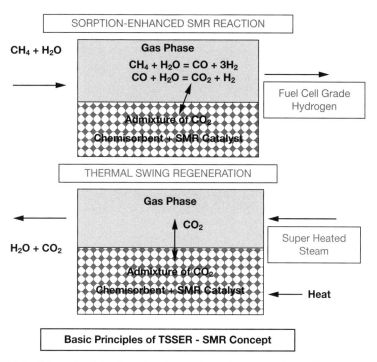

FIGURE 2.1   Schematic of the thermal swing sorption-enhanced reaction (TSSER)-steam-methane reforming (SMR) concept. *Source*: Reproduced with permission of Beaver et al. [2].

In a recent study, a new concept based on sorption-enhanced reaction (SER) for steam-methane reforming (SMR) was demonstrated to directly produce fuel-cell grade H_2 (~10 ppm CO) with very high CH_4 to H_2 conversion efficiency (>99%) using an admixture of a commercial SMR catalyst and a CO_2 chemisorbent, K_2CO_3 promoted hydrotalcite [2]. The reaction was carried out at a much lower temperature (520–590°C) than the conventional reaction temperature of 700–900°C without sacrificing the reactor performance. It also eliminates the subsequent H_2 purification step by a conventional PSA process. Figure 2.1 shows a schematic of the thermal swing sorption-enhanced reaction (TSSER)-steaming methane reforming (SMR) concept. The overall outcomes are attributed to four related factors: (a) favorable thermodynamic equilibrium of the highly endothermic SMR reaction at the higher reaction temperature, (b) faster kinetics of SMR reaction at higher temperatures, (c) favorable removal of CO_2 from the reaction zone at lower temperatures, and (d) higher cyclic working capacity for CO_2 chemisorption at higher temperature.

Usually, following reforming, the resulting syngas is sent to one or more shift reactors in which the water-gas shift reaction takes place and generates more hydrogen:

$$CO + H_2O \rightarrow CO_2 + H_2. \tag{2.2}$$

In practice, this shift reaction is often conducted in two stages: first in a high temperature shift reactor operating in the 350–475°C range that does most of the conversion and then a low temperature reactor operating in the 200–250°C range that brings the CO concentration to less than a few % by volume. Lower temperature reaction requires more active catalysts.

The final step is hydrogen purification, which can be accomplished using PSA systems or palladium membranes, producing hydrogen with up to 99.999% purity. CO removal is a major step and can be done using preferential oxidation based on the following reaction:

$$CO + \tfrac{1}{2} O_2 = CO_2. \tag{2.3}$$

This reaction is strongly favored over hydrogen oxidation, thereby allowing preferential removal of CO to the level of a few part per million.

Methane steam reformers have been built over a large range of sizes and types, including conventional, compact "fuel cell type," plate-type, and membrane reactors, as summarized in an excellent review article [1]. We will not get into details of reformers here. Briefly, Figure 2.2 shows a schematic of a small-scale methane steam reformer designed for fuel cell applications that involved convective heat transfer. Compared with conventional tube reformers, such "fuel cell type" reformers can operate at lower pressure and temperature with lower cost materials, reducing the overall cost associated with the operation. This type of reformers is commercially available.

2.3   PARTIAL OXIDATION

Another important method for H_2 production is partial oxidation of hydrocarbons. Figure 2.3 shows a flow chart on the comparison between the steam reforming and partial oxidation methods [1]. While both methods are based on thermochemistry for hydrogen generation, they differ mainly in the source of oxygen, that is, water for steam reforming and oxygen gas for partial oxidation.

For example, methane can be partially oxidized to produce CO and H_2 based on the following reaction:

$$2\,CH_4 + O_2 \rightarrow 2\,CO + 4\,H_2. \tag{2.4}$$

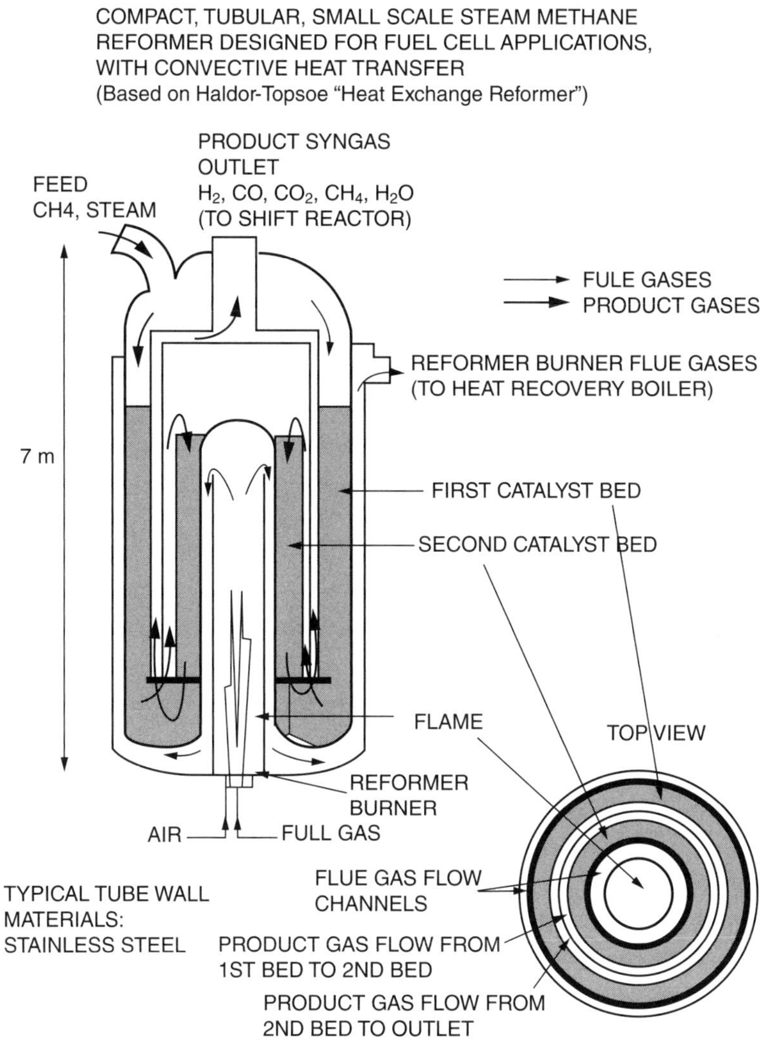

COMPACT, TUBULAR, SMALL SCALE STEAM METHANE
REFORMER DESIGNED FOR FUEL CELL APPLICATIONS,
WITH CONVECTIVE HEAT TRANSFER
(Based on Haldor-Topsoe "Heat Exchange Reformer")

FIGURE 2.2 Schematic of a compact, tubular, small-scale methane reformer designed for fuel cell applications with convective heat transfer. *Source*: Reproduced with permission from Ogden [1].

The standard heat of reaction of this reaction is $\Delta H_{298} = -71$ kJ. Since it is exothermic, no indirect heat exchanger is needed. Because of the high temperature involved, no catalysts are required for the reaction. However, the hydrogen yield can be substantially increased with the use of catalysts. Compared with Equation 2.1, this reaction is clearly lower in yield of H_2 for a given amount of CH_4. A typical hydrogen production plant based on partial oxidation includes a partial oxidation reactor, a shift reactor, and hydrogen purification equipment.

THERMOCHEMICAL HYDROGEN PRODUCTION METHODS

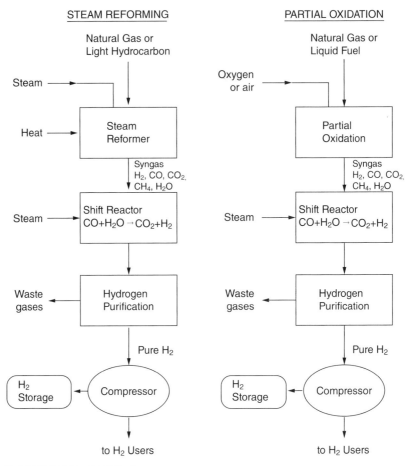

FIGURE 2.3   Comparison of the steam reforming and partial oxidation methods. *Source:* Reproduced with permission from Ogden [1].

2.4   METHANOL AND ETHANOL STEAM REFORMING

Besides methane, methanol is another major candidate for H_2 production via steam reforming through the following reaction:

$$CH_3OH + H_2O \rightarrow CO_2 + 3 H_2 \quad \Delta H = +49 \text{ kJ} \cdot \text{mol}^{-1}. \quad (2.5)$$

Among the different possible reactions involving methanol for hydrogen generation, this reaction offers the highest maximum hydrogen content in the product gas (75%) [3]. In addition, since no gases need to be compressed in the feed, the reaction can easily be carried out at higher pressure, thus keeping membranes as an option for the successive gas clean-up step. The endothermicity of the reaction, however, requires external heating of the reactor, which makes it challenging to achieve short start-up times and fast transient behavior desired for some applications, such as in automobiles. Like in steam methane reforming, efficient conversion of methanol to hydrogen requires the use of catalysts. Common catalysts for this purpose include $Cu/ZnO/Al_2O_3$, $Cu-Mn-O$, and Cu-based binary metals, such as Cu/Cr, Cu/Zn, and Cu/Zr [4, 5]. The choice of catalyst has a great influence on the methanol conversion and carbon dioxide selectivity of the reforming reaction.

Besides large-scale reformers, there is interest in developing small reformers for portable electronics applications. Toward this goal, a silicon-chip based microreactor has been successfully fabricated and tested to carry out the reaction of methanol reforming for microscale hydrogen production [6]. The developed microreactor, in conjunction with a micro fuel cell, is proposed as an alternative to conventional portable sources of electricity, such as batteries, due to its ability to provide an uninterrupted supply of electricity as long as a supply of methanol and water can be provided. The microreformer-fuel cell combination does not require recharging, as compared with conventional rechargeable lithium-ion batteries, and affords a significantly higher energy storage density. The microreactor consists of a network of catalyst-packed parallel microchannels of depths ranging from 200 to 400 μm with a catalyst particle filter near the outlet fabricated using photolithography and deep-reactive ion etching (DRIE) on a silicon substrate. Experimental runs have demonstrated a methanol to hydrogen molar conversion of at least 85–90% at flow rates enough to supply hydrogen to an 8- to 10-W fuel cell.

Similar to methane or methanol, ethanol can also be used for hydrogen generation through steam reforming. Unlike methane and methanol, however, ethanol can be prepared from agricultural products and residues, and thus represents a renewable resource. One of the major mechanisms proposed for ethanol steam reforming is given below [7]:

$$C_2H_5OH + H_2O \rightarrow CH_4 + CO_2 + 2\,H_2 \qquad (2.6)$$

$$CH_4 + H_2O \rightarrow CO + 3\,H_2 \qquad (2.7)$$

$$CO + H_2O \rightarrow CO_2 + H_2. \qquad (2.8)$$

Similar to other steam reforming reactions, catalysts play a critical role in ethanol steam reforming [7]. For example, an earlier study has found that

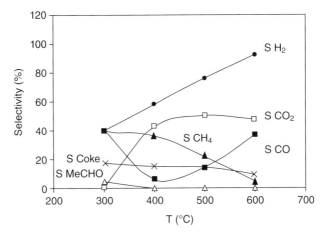

FIGURE 2.4   Selectivities of ethanol reforming as a function of temperature (H2O/EtOH = 3.7). *Source*: Reproduced with permission from Fierro et al. [9].

Ni/La$_2$O$_3$ catalyst exhibits high activity and good long-term stability for steam reforming of ethanol and is a good candidate for ethanol reforming processors for fuel cell applications [8]. Another study has found that oxidative reforming of biomass-derived ethanol can be carried out over an inexpensive Ni–Cu/SiO$_2$ catalyst with respect to solid polymer fuel cell (SPFC) applications [9]. The reaction can be performed either under diluted conditions (with helium as diluent) or under conditions corresponding to an on-board reformer. Selectivity of ethanol reforming depends on a number of operating parameters, including reaction temperature, H$_2$O/EtOH molar ratio, and O$_2$/EtOH molar ratio of the feed to the reformer. Figure 2.4 shows the effect of the reaction temperature on the selectivity of the reforming reaction. The hydrogen content and the CO$_2$/CO$_x$ molar ratio in the outlet gases have been used as parameters to optimize the operating conditions in the reforming reactor. The tests carried out at on-board reformer conditions have shown that an H$_2$O/EtOH molar ratio = 1.6 and an O$_2$/EtOH molar ratio = 0.68 at 973 K allow a hydrogen-rich mixture (33%) considered of high interest for SPFC. The use of oxygen decreases the production of methane and coke that, in turn, increases the lifetime of the catalyst, which has been demonstrated to exhibit good long-term stability.

In a recent study, dielectric barrier discharge (DBD), the electrical discharge between two electrodes separated by an insulating dielectric barrier, was used for the generation of hydrogen from ethanol reforming [10]. It was found that the increase of ethanol flow rate decreased ethanol conversion efficiency and hydrogen yield, and high water/ethanol ratio and addition of oxygen were advantageous for hydrogen production. Figure 2.5 shows the

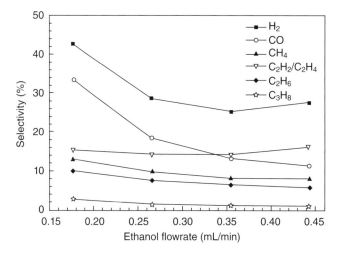

FIGURE 2.5 Dependence of selectivity of product gases on the flow rate of ethanol. *Source*: Reproduced with permission from Wang et al. [10].

dependence of selectivity of product gases on the flow rate of ethanol. The ethanol conversion efficiency and hydrogen yield increased with the vaporization at room temperature up to the maximum at first, and then decreased slightly. The maximum hydrogen yield of 31.8% was obtained at an ethanol conversion of 88.4% under the optimum operation conditions of vaporization temperature of 120°C, ethanol flux of 0.18 mL min^{-1}, water/ethanol ratio of 7.7, and oxygen volume concentration of 13.3%.

2.5   GLYCEROL REFORMING

Glycerol represents another important renewable source for H_2. With a chemical formula of $C_3H_8O_3$, it is a saturated and oxygenated hydrocarbon with one OH group on each of the three carbon atoms and a one-to-one oxygen-to-carbon ratio. It is edible, biodegradable, nonflammable, nontoxic, and high boiling. Glycerol can be synthesized from propylene oxide, sorbitol, or glucose, or obtained as a byproduct in several industrial processes, such as soap manufacturing, biodiesel production, or lignocellulose-to-ethanol conversion.

2.5.1   Glycerol Reforming Processes

Glycerol reforming has been extensively studied recently, including steam or aqueous reforming, catalytic partial oxidation, and autothermal reforming

[11]. The different reforming processes are classified based on the source of heat and the type of reactants. In general, the glycerol reforming is endothermic and can be represented as follows:

$$C_3H_8O_3 + x\,H_2O + yO_2 \rightarrow a\,CO_2 + b\,CO + c\,H_2O + d\,H_2 + e\,CH_4. \quad (2.9)$$

When $x > 0$ and $y = 0$, the process is steam reforming that, as a vapor phase catalytic reaction, occurs at high temperature (e.g., 800°C) and low pressure (e.g., 0.1 MPa). The reaction can be represented as:

$$C_3H_8O_3 + 3\,H_2O \rightarrow 3\,CO_2 + 7\,H_2 \quad \Delta H = 128\;kJ\cdot mol^{-1}. \quad (2.10)$$

For every mole of glycerol, 7 mol of H_2 is produced based on stoichiometry, which is highly favorable. Major issues include byproduct formation (e.g., CO), catalysts deactivation, and high energy consumption. To lower the energy requirement, the reforming can be carried out in the aqueous phase at lower temperature (e.g., 225°C) and high pressure (e.g., 2 MPa). One added advantage of this process is the lower CO concentration in the reformate since the water gas shift reaction is favored at lower temperature. However, limitations of this process include the high pressure requirement and low H_2 selectivity due to the possible formation of alkanes.

When $x = 0$ and $y > 0$ and oxygen is used to provide the heat required internally, catalytic partial oxidation takes places and can be represented as:

$$C_3H_8O_3 + 3/2\,O_2 \rightarrow 3\,CO_2 + 4\,H_2 \quad \Delta H = -603\;kJ\cdot mol^{-1}. \quad (2.11)$$

This overall process is exothermic and produces 4 mol of H_2 for every mole of glycerol. The conversion efficiency, as well as reaction temperature and pressure required, depend strongly on the catalysts used, for example, Ni, Co, Pt, Ru, and Rh. A more detailed discussion of some examples of catalysts and related reactions mechanisms will be given later.

Similarly, in autothermal reforming ($x > 0$ and $y > 0$), oxygen, water steam, and glycerol react in the presence of a catalyst. The reaction can be given as:

$$C_3H_8O_3 + 3/2\,H_2O + \tfrac{3}{4}\,O_2 \rightarrow 3\,CO_2 + 11/2\,H_2 \quad \Delta H = -240\;kJ\cdot mol^{-1}. \quad (2.12)$$

For every mole of glycerol, 5.5 mol of H_2 is produced. It should be pointed out that the heat of reactions for processes represented by Equation 2.10, Equation 2.11, and Equation 2.12 vary substantially in different reports [12].

2.5.2   Mechanistic Aspects of Glycerol Reforming Reactions

The overall reforming process involves a number of complex reactions, some of which lead to the formation of byproducts. The primary reactions are given by [13]:

$$C_3H_8O_3 \rightarrow 3\,CO + 4\,H_2 \quad \Delta H = 251\,kJ \cdot mol^{-1} \tag{2.13}$$

$$CO + H_2O \rightarrow CO_2 + H_2 \quad \Delta H = -41\,kJ \cdot mol^{-1}. \tag{2.14}$$

Equation 2.13 represents glycerol decomposition in the presence of water, whereas as Equation 2.14 denotes the water gas shift reaction. The sum of the two reactions is the overall reaction represented in Equation 2.10. No oxygen is required for this reaction. High temperatures, low pressures, and high water/glycerol ratios favor H_2 production.

Some of the side reactions, including methane formation, are represented below:

$$CO + 3\,H_2 \rightarrow CH_4 + H_2O \quad \Delta H = -206\,kJ \cdot mol^{-1} \tag{2.15}$$

$$CO_2 + 4\,H_2 \rightarrow CH_4 + 2\,H_2O \quad \Delta H = -165\,kJ \cdot mol^{-1}. \tag{2.16}$$

The formation of CH_4 is favored at high pressures, low temperatures, and low water/glycerol ratios. The methane formed can react with CO_2 to produce CO and H_2. Other side-reactions can lead to coke formation on the surface of catalysts, and some of which are represented by the following reactions:

$$2\,CO \rightarrow CO_2 + C \quad \Delta H = -171.5\,kJ \cdot mol^{-1} \tag{2.17}$$

$$CO + H_2 \rightarrow H_2O + C \quad \Delta H = -131\,kJ \cdot mol^{-1}. \tag{2.18}$$

In addition, CH_4 decomposition and reaction between CO_2 and H_2 can also result in coke formation. To minimize methane and coke formation and enhance H_2 production, the reactions are usually carried out at high temperature ($>627°C$), a pressure of 0.1 MPa, and a water/glycerol molar ratio of 9 [11].

Similar mechanistic and thermodynamic studies have been conducted for other reforming processes, for example, those involving aqueous water and H_2O_2. Oxidation plays an important role in these cases.

2.5.3   Catalytic Reforming of Glycerol

Similar to reforming of methane and methanol or ethanol, catalysts can accelerate the glycerol reforming process. Catalysts such as Ni, Co, Pt, Ru, and Rh, supported by metal oxides such as Al_2O_3, TiO_2, MgO, or CeO_2, are often used. The catalytic reactions are proposed to involve dehydrogenation of the glycerol and chemisorption on the catalyst surface via carbon and/or oxygen atoms. Cleavage of C–C bond ($\Delta H = 347$ kJ mol^{-1}) and subsequent dehydrogenation result in the formation of adsorbed CO, which is desired for reactions such as the water–gas shift reaction. However, C–O bond cleavage ($\Delta H = 358$ kJ mol^{-1}) and subsequent hydrogenation can occur and lead to formation of side-products, such as small alkanes and alcohols.

The outcome of the reforming process depends on a number of important factors, including the chemical nature of the catalyst, temperature, pressure, and initial water/glycerol molar ratio. For a given temperature (e.g., 900°C) and water/glycerol ratio (e.g., 9:1), the glycerol conversion has been found to be in the order of Ni > Ir > Pd > Rh > Pt > Ru, while the H_2 selectivity is in the order of Ni > Ir > Ru > Pt > Rh, Pd [14]. For a given catalyst, the reaction temperature and water/glycerol ratio are critical in determining the glycerol conversion efficiency and hydrogen selectivity. Figure 2.6 shows an example of the dependence of hydrogen selectivity and glycerol conversion efficiency on the catalyst used, Ni/Al_2O_3 versus $Rh/CeO_2/Al_2O_3$ [14]. While $Rh/CeO_2/Al_2O_3$ showed better glycerol conversion efficiency, Ni/Al_2O_3 exhibits much better hydrogen selectivity.

2.6 CRACKING OF AMMONIA AND METHANE

Another approach to hydrogen generation is thermal decomposition or cracking of inorganic or organic compounds that contain hydrogen. Examples of such compounds include ammonia and methane.

2.6.1 Ammonia Cracking

Ammonia is an important industrial chemical produced in large quality each year. It is used for many applications, including fertilizers in agriculture. Since it can be relatively easily stored and transported, and contains a significant amount of hydrogen, it is considered as a good source for hydrogen. One additional advantage is that N_2 as the by-product of hydrogen generation can be released into the atmosphere without causing any negative environmental impact. The hydrogen generation reaction is as follow:

$$2\,NH_3 \rightarrow N_2 + 3\,H_2 \quad \Delta H = 46.22 \text{ kJ} \cdot \text{mol}^{-1}. \quad (2.19)$$

It is usually conducted at high temperature (>900°C) in a process called cracking [15]. The cracking process is thermally efficient and simple. It is

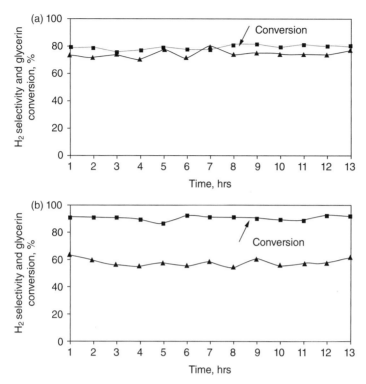

FIGURE 2.6   Hydrogen selectivity and glycerol conversion over (a) Ni/Al_2O_3 and (b) $Rh/CeO_2/Al_2O_3$ for 13 hours at 900°C, with a feed flow rate of 0.15 mL min^{-1} and water to glycerol ratio of 6. *Source*: Reproduced with permission from Adhikari et al. [14].

usually not necessary to carry out fine purification after ammonia cracking. Furthermore, co-reactants such as water are not required.

The reaction rate depends on temperature, pressure, and the catalyst used. The theoretical limit for the lowest working temperature possible is given by the chemical equilibrium for the dissociation reaction. Figure 2.7 shows that a nearly complete conversion from ammonia to hydrogen and nitrogen at higher temperatures (near 430°C) and atmospheric pressure is possible [15].

Catalysts used for ammonia dissociation include materials such as porcelain or silica glass, metals such as Fe, W, Mo, and Ni, as well as noble metals and metal oxides. Active catalytic reactions are usually conducted in the temperature range from 700 to 1100°C. Commercially available simple catalyst materials such as nickel oxide or iron oxide on aluminum and the influence of the addition of noble metals such as Pt have been investigated.

The synthesis of ammonia in industry is usually based on the reverse process of Equation 2.19, that is reaction between hydrogen and nitrogen,

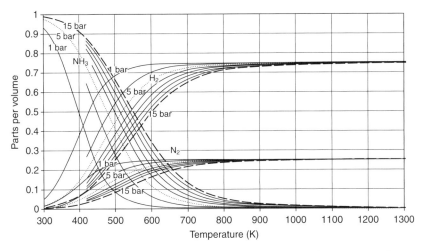

FIGURE 2.7   Chemical equilibrium of $2NH_3 = N_2 + 3 H_2$ as a function of temperature and pressure. *Source*: Reproduced with permission from Hacker and Kordesch [15].

which is exothermic thermodynamically but is only effective under pressure and high temperature and with catalysts due to high kinetic barriers. Also, ammonia cannot be simply released since it is hazardous. Thus, it is not easy to use the reaction between hydrogen and nitrogen as a direct means for energy generation as in the case of hydrogen reaction with oxygen. However, if ammonia can be produced at a low cost, it is a potential source for hydrogen, which is a promising energy carrier.

One major driving force for ammonia synthesis is nitrogen fixation for producing fertilizers. There are a number of large-scale ammonia production plants worldwide, producing a total of 131 million tons of ammonia in 2010 and projected to reach near 200 million tons by 2012. Hydrogen is used for ammonia synthesis and is mainly produced from natural gas or other liquefied petroleum gases such as propane or butane, or petroleum naphtha. Figure 2.8 shows a flow chart of typical options for producing and purifying ammonia synthesis gas.

The first step in the process is to remove sulfur compounds from the feedstock because sulfur deactivates the catalysts used in subsequent steps. Sulfur removal requires catalytic hydrogenation to convert sulfur compounds to gaseous hydrogen sulfide, which is then absorbed and removed by passing through beds of ZnO, where it is converted to solid zinc sulfide. The second step is catalytic steam reforming of the sulfur-free feedstock to form hydrogen plus carbon monoxide. The third step uses catalytic shift conversion to convert the carbon monoxide (reacting with H_2O) to carbon dioxide (which

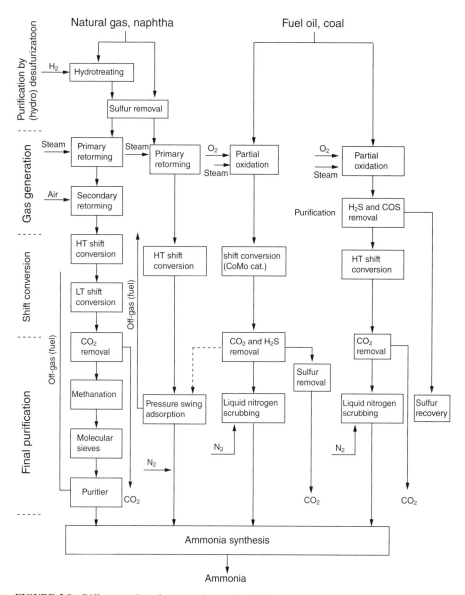

FIGURE 2.8 Different options for generating and purifying ammonia synthesis gas. *Source*: Reproduced with permission from Appl [16].

is then removed) and hydrogen. The fourth step is production of hydrogen via catalytic methanation, including removal of any small residual amounts of carbon monoxide or carbon dioxide from the hydrogen. The final step is to produce the desired end-product ammonia, in which hydrogen is catalytically reacted with nitrogen (derived from process air) to form anhydrous liquid ammonia. This step is known as the ammonia synthesis loop (also referred to as the Haber–Bosch process).

2.6.2 Methane Cracking

Similarly, hydrogen can be produced by cracking of methane. Compared with methane steam reforming, no CO is generated during the cracking reaction, thus there is no need for separating CO from hydrogen after the reaction. The methane cracking reaction is as follows:

$$CH_4 = C_s + 2H_2 \quad H = 74.8 \text{ kJ} \cdot \text{mol}^{-1}, \tag{2.20}$$

where C_s stands for solid carbon. Besides the desired product, hydrogen, the only byproduct is carbon, which is usually in the form of filamentous carbon or carbon nanotubes [17]. Separation of unreacted methane and hydrogen can be readily achieved by adsorption or membrane separation to produce a stream of hydrogen with 99% by volume, which is much simpler than the reforming process with complicated separation processes that involve CO_2 or CO. This can be especially important for proton-exchange membrane (PEM) fuel cells, in which the Pt-based electrocatalyst can be poisoned by CO. The carbon nanotubes produced as a solid product are commercially useful in applications such as adsorption, catalysis, or carbon storage.

The cracking reaction is strongly temperature dependent and much more effective with catalysts used. Examples of catalysts include Ni, Fe, Co, and activated carbon, usually supported by substrates like SiO_2, TiO_2, ZrO_2, Al_2O_3, MgO, graphite, or composites of the oxides. Noncatalytic methane cracking is very slow at temperatures below 1000°C, while catalytic cracking of methane can be conducted at temperatures as low as 500°C. Figure 2.9 shows the predicted equilibrium methane conversion as a function of temperature based on thermodynamic considerations and data without using catalyst [17]. Also shown in the figure is the number of moles of CH_4, H_2, and C for an initial 100 mol of CH_4. It is clear that the equilibrium conversion increases with increasing temperature, starting from 30% conversion at 500°C to almost complete conversion at 1000°C.

For catalytic reactions, iron group metals are known to have the highest activity for general hydrocarbon cracking. For methane, which is the most stable compared with other hydrocarbons, Ni has been found to be the most active catalyst among the iron group metals. Direct comparison shows that the catalytic activity for the iron group metals is: Ni > Co > Fe [18]. The byproduct, carbon, can cause deactivation of the catalysts by encapsulating them. Regeneration of the catalysts can be done with steam or air.

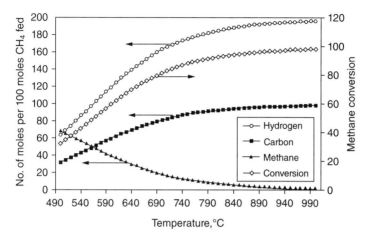

FIGURE 2.9 Equilibrium composition and conversion as a function of temperature. The figure shows the equilibrium number of moles based on an initial 100 mol of CH_4. *Source*: Reproduced with permission from Amin et al. [17].

Different mechanisms of methane cracking reaction have been proposed [17, 19, 20]. Most of the proposed mechanisms involve step-wise removal of H from the initial CH_4, via radicals such as CH_3, CH_2, and CH, and finally formation of H_2 from H atoms. The initial CH_4 and the reaction intermediates or radicals are adsorbed on the catalysts that lower the activation energy of reactions, leading to breaking of the C–H bonds. The initial CH_4 can be dissociatively or nondissociatively adsorbed while the subsequent steps are similar.

2.6.3 Other Decomposition Methods

There are other ways for generating hydrogen from hydrocarbons. For example, methane and higher hydrocarbons can be directly decomposed to hydrogen and carbon over molten magnesium (Mg) as a catalyst [21]. Hydrogen as well as micrometer-size carbon particles could be formed as products. The catalyst loses activity because of the evaporation of metal Mg. The activity of the catalyst can be recovered by heating the upper cold section of the reactor to circulate Mg back to the reactor bed. Mg_2C_3 was has been identified as reaction intermediate in the reaction, as shown in the proposed reaction mechanism in Figure 2.10. It was suggested that since magnesium is more electropositive than carbon and hydrogen, the breaking of C–C and C–H bonds should involve the valence electrons of magnesium. With the evolution of hydrogen gas and the decomposition of the magnesium carbides, metallic magnesium is regenerated.

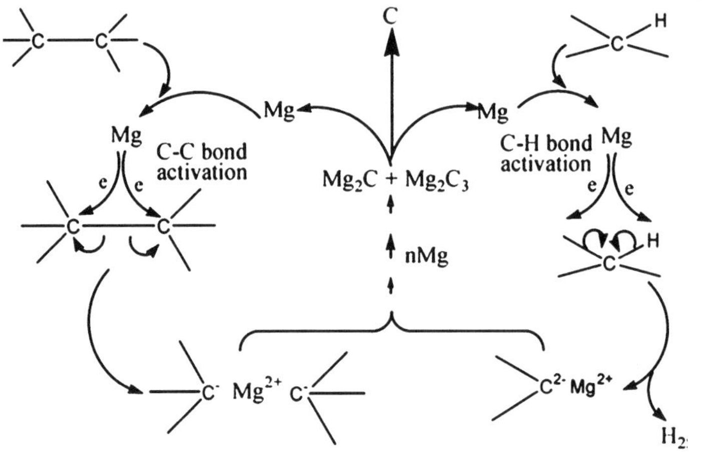

FIGURE 2.10   Proposed mechanism of reaction. *Source*: Reproduced from Wang et al. [21].

Compared with the reforming process, the hydrocarbon direct decomposition process produces CO-free hydrogen, does not emit CO_2 to environment, generates useful carbon powder, and is also an energy-saving approach for hydrogen preparation. To form each mole of hydrogen by methane, direct decomposition only consumes about 65.1% of the energy as that needed in the steam reforming of methane. The reaction is also proved to be useful in the decomposition of waste polyolefins, such as poly olefin plastic and rubber.

2.7   SUMMARY

Hydrocarbons are important carriers and sources of hydrogen. Due to their high chemical stability, generation of hydrogen from hydrocarbons is generally an energy-consuming process, and usually requires heating and catalysis for high efficiency. Compared with other methods of hydrogen generation, using hydrocarbons has some pros and cons. The main limitation is that it uses fossil-based chemicals. One major advantage is that the major infrastructure is already in existence for handling and delivering hydrocarbons. At least for the foreseeable future, hydrocarbons will still be expected to be the major source for hydrogen. The need for small-scale reformers for portable or mobile applications is expected to grow, and further research and development are needed in these areas.

REFERENCES

1. Ogden, J. *Review of Small Stationary Reformers for Hydrogen Production*, The International Energy Agency, Paris, 2001.

2. Beaver, M.G., Caram, H.S., Sircar, S. Sorption enhanced reaction process for direct production of fuel-cell grade hydrogen by low temperature catalytic steam-methane reforming. *Journal of Power Sources* **2010**, *195*(7), 1998–2002.

3. Geissler, K., Newson, E., Vogel, F., Truong, T.B., Hottinger, P., Wokaun, A. Autothermal methanol reforming for hydrogen production in fuel cell applications. *Physical Chemistry Chemical Physics* **2001**, *3*(3), 289–293.

4. Papavasiliou, J., Avgouropoulos, G., Ioannides, T. In situ combustion synthesis of structured Cu–Ce–O and Cu–Mn–O catalysts for the production and purification of hydrogen. *Applied Catalysis B-Environmental* **2006**, *66*(3–4), 168–174.

5. Lindstrom, B., Agrell, J., Pettersson, L.J. Combined methanol reforming for hydrogen generation over monolithic catalysts. *Chemical Engineering Journal* **2003**, *93*(1), 91–101.

6. Pattekar, A.V., Kothare, M.V. A microreactor for hydrogen production in micro fuel cell applications. *Journal of Microelectromechanical Systems* **2004**, *13*(1), 7–18.

7. Haryanto, A., Fernando, S., Murali, N.; Adhikari, S. Current status of hydrogen production techniques by steam reforming of ethanol: A review. *Energy & Fuels* **2005**, *19*(5), 2098–2106.

8. Fatsikostas, A.N., Kondarides, D.I., Verykios, X.E. Steam reforming of biomass-derived ethanol for the production of hydrogen for fuel cell applications. *Chemical Communications* **2001**, (9), 851–852.

9. Fierro, V., Klouz, V., Akdim, O., Mirodatos, C. Oxidative reforming of biomass derived ethanol for hydrogen production in fuel cell applications. *Catalysis Today* **2002**, *75*(1–4), 141–144.

10. Wang, B.W., Lu, Y.J., Zhang, X., Hu, S.H. Hydrogen generation from steam reforming of ethanol in dielectric barrier discharge. *Journal of Natural Gas Chemistry* **2011**, *20*(2), 151–154.

11. Vaidya, P.D., Rodrigues, A.E. Glycerol reforming for hydrogen production: A review. *Chemical Engineering & Technology* **2009**, *32*(10), 1463–1469.

12. Dauenhauer, P. J., Salge, J. R. Schmidt, L. D., Renewable hydrogen by autothermal steam reforming of volatile carbohydrates. *Journal of Catalysis* **2006**, *244*(2), 238–247.

13. Sutton, D., Kelleher, B., Ross, J.R.H. Review of literature on catalysts for biomass gasification. *Fuel Processing Technology* **2001**, *73*(3), 155–173.

14. Adhikari, S., Fernando, S., Haryanto, A. Production of hydrogen by steam reforming of glycerin over alumina-supported metal catalysts. *Catalysis Today* **2007**, *129*(3–4), 355–364.

15. Hacker, V., Kordesch, K. Ammonia crackers. In W. Vielstich, A. Lamm, and H.A. Gasteiger, eds., *Handbook of Fuel Cells: Fundamentals, Technology and Applications*, John Wiley & Sons, Chichester, 2003, Vol. 3, pp. 121–127.

16. Appl, M. *Ammonia, Principles and Industrial Practice*. New York, Wiley-VCH, 1999.

17. Amin, A.M., Croiset, E.M., Epling, W. Review of methane catalytic cracking for hydrogen production. *International Journal of Hydrogen Energy* **2011**, *36*(4), 2904–2935.

18. Avdeeva, L.B., Reshetenko, T.V., Ismagilov, Z.R., Likholobov, V.A. Iron-containing catalysts of methane decomposition: Accumulation of filamentous carbon. *Applied Catalysis a-General* **2002**, *228*(1–2), 53–63.

19. Grabke, H.J. Evidence on the surface concentration of carbon on gamma iron from kinetics of carburization in CH_4–H_2. *Metallurgical Transactions* **1970**, *1*(10), 2972–2975.

20. Alstrup, I., Tavares, M.T. The kinetics of carbon formation from $CH_4 + H_2$ on a silica-supported nickel-catalyst. *Journal of Catalysis* **1992**, *135*(1), 147–155.

21. Wang, K., Li, W.S., Zhou, X.P. Hydrogen generation by direct decomposition of hydrocarbons over molten magnesium. *Journal of Molecular Catalysis A: Chemical* **2008**, *283*(1–2), 153–157.

3

Solar Hydrogen Generation: Photocatalytic and Photoelectrochemical Methods

3.1 BASICS ABOUT SOLAR WATER SPLITTING

Water splitting is a chemical process that converts water into hydrogen and oxygen. It represents one of the most important reaction for hydrogen fuel production, as water is the most abundant hydrogen source on the Earth. However, water splitting is a thermodynamically uphill reaction:

$$H_2O \rightarrow H_2 + \frac{1}{2}O_2 \quad (\Delta G \sim 237.2 \ kJ \cdot mol^{-1}; 1.23 \ V \ vs. \ NHE).$$

Therefore, a minimum potential of 1.23 V is required to overcome this thermodynamic barrier. In practice, there is an additional kinetic barrier (overpotential) for charge carrier transfer for water oxidation and proton reduction. The overpotentials for these reactions are material dependent. Thus, practically the minimum energy required for water splitting is higher than 1.23 V. The required energy can be obtained from a nonrenewable or renewable energy sources. For instance, electrolysis is the most direct method to split water by applying a potential higher than the minimum required energy. The common industrial electrolyzer with platinum as catalyst can achieve a hydrogen production efficiency of around 70% [1]. Nevertheless,

Hydrogen Generation, Storage, and Utilization, First Edition. Jin Zhong Zhang, Jinghong Li, Yat Li, and Yiping Zhao.
© 2014 John Wiley & Sons, Inc. Published 2014 by John Wiley & Sons, Inc.

this process simply transforms electricity into chemical energy in the form of hydrogen, and electric power expense has the largest share in the price of electrolytic hydrogen, which is not a promising solution for energy sustainability. Additionally, using platinum electrodes significantly increases the cost of electrolyzers, and thus, the cost of producing hydrogen.

Alternatively, a number of methods have been developed to split water in a clean and more cost-effective manner by using renewable solar energy [2]. For instance, an electrolyzer system can be powered by solar cells, such as silicon-based solar cells or dye-sensitized solar cells. The solar cells can harvest solar energy and provide photovoltages. Therefore, solar cells can be connected in series to supply the required potential for electrolysis. Using a combination of conventional electrolyzer and a commercially available solar cell with 10–15% of conversion efficiency, a solar-to-hydrogen efficiency of ~10% could be potentially achieved. Nevertheless, the relatively high cost of solar cells and electrolyzers are major drawback of this approach. In this regard, a photocatalytic or photoelectrochemical (PEC) system consisting of semiconductor materials that can harvest light and use this energy directly for splitting water is a more promising and cost-effective way for solar hydrogen generation. In this chapter, we will highlight recent research progress in photocatalytic and PEC water splitting.

3.2 PHOTOCATALYIC METHODS

3.2.1 Background

With the rapid growth of global population, new energy resources should be explored to address the continuously growing demand for energy. Renewable solar energy is a promising solution to energy sustainability. In order to obtain continuous and stable power supply, it requires efficient and cost effective energy storage carriers or device to store solar energy during intermittent sunlight irradiation [3]. Inspired by photosynthesis, which converts solar energy into chemical energy that is stored in the form of carbohydrates, great efforts have been made to mimic this natural process using man-made materials for solar water splitting to generate hydrogen fuel [3, 4]. Ideally, hydrogen gas can be continuously generated when photocatalyst powders are dispersed in water under solar light illumination. However, the photocatalyst must overcome the uphill Gibbs free energy change to drive this reaction [5].

The first demonstration of artificial photosynthesis was reported by Honda and Fujishima in 1972 using semiconductor TiO_2 as photocatalyst [6]. When the energy of incident light is larger than the bandgap energy of the semiconductors, photoexcited electrons and holes will be created in the conduction band and valence band, respectively. Water molecules will be reduced by the electrons in the conduction band to form H_2 gas, and oxidized by holes in the valence band to generate O_2 gas [5]. To serve as good photocatalysts, semiconductor materials should satisfy several key requirements [7]. First, the band edge positions should straddle the electrolysis potentials of H^+/H_2 (0 V vs. NHE) and O_2/H_2O (1.23 V vs. NHE). Therefore, the theoretical minimum energy required for water splitting is 1.23 eV. However, taking account of the presence of overpotential for water reduction and oxidation on semiconductor surfaces, the actual energy for water splitting should be at least 1.8 eV [5]. Second, the bandgap of photocatalysts should be small enough to absorb a significant portion of solar light. Third, the semiconductors must be chemically stable during water oxidation and reduction in aqueous solution [7].

A number of strategies have been developed to improve the photocatalytic water-splitting performance of semiconductors. One strategy is bandgap engineering to improve photon absorption efficiency. The overall solar-to-hydrogen efficiency is closely related to the light absorption efficiency. For example, nitrogen doping has been demonstrated to increase visible light photoactivity of TiO_2 in water splitting by introducing nitrogen as impurities to narrow the bandgap [8]. A second strategy is improvement of charge separation and suppression of the electron–hole recombination loss. For instance, considerable efforts have been placed to develop semiconductor nanostructures that have a large surface area and short carrier diffusion length, which are beneficial for charge separation and suppression of electron–hole recombination [7]. A third strategy is the construction of surface reaction sites for water oxidation or reduction. For example, Pt-modified semiconductors have been used to reduce the overpotential of water reduction and to improve the efficiency of hydrogen generation [9].

A number of semiconductors have been studied for solar hydrogen generation. These semiconductors can be classified into metal oxides such as TiO_2 [10, 11], ZnO [12, 13], and $SrTiO_3$ [14], metal oxynitrides such as TaON [15, 16], metal nitrides/phosphide such as Ta_3N_5 and InP_3 [17, 18], metal chalcogenides [9, 12], and silicon [19, 20]. The conduction bands of all these semiconductors are more negative than water reduction potential (0 vs. NHE), which ensure spontaneous water reduction when the semiconductors are illuminated with light energy larger than their bandgap energies. In this chapter, we will discuss the recent progress in developing different kinds of semiconductors for photocatalytic hydrogen generation.

3.2.2   Metal Oxides

Metal oxides have been extensively studied as photocatalysts for solar hydrogen generation, since the first demonstration of photocatalytic study of TiO_2 in 1972 [6]. TiO_2 and ZnO are two common semiconductor photo-catalysts, due to their low cost and favorable band edge positions for hydro-gen generation [10, 11, 14]. However, the photocatalytic performance of ZnO and TiO_2 are limited by their large bandgap and rapid charge recombi-nation. In this regard, nanostructured metal oxides have been designed to improve the charge separation at the interface between semiconductor and electrolyte [4, 7]. Element doping, such as nitrogen doping, has also been used to narrow their bandgap and thereby increase the visible light absorp-tion [8, 13, 21].

Recently, Chen et al. developed an alternative approach to improve visible and near infrared optical absorption of TiO_2 by introducing surface disorder through hydrogen treatment [22]. Hydrogen-treated TiO_2 nanoparticle con-sists of a crystalline TiO_2 core and a highly disordered surface layer [22]. Creation of large amount of lattice disorder yields mid-gap states in TiO_2. Instead of forming discrete donor states near the conduction band edge, these mid-gap states can form a continuum extending to and overlapping with the conduction band edge, which are often also known as band tail states [23]. Similarly, the large amount of disorder can result in band tail states merging with the valence band [23]. Therefore, the bandgap and optical property of TiO_2 can be significantly changed by creating surface disorder. The color of TiO_2 changed from white to black, indicating TiO_2 is able to absorb visible light. Notably, the black TiO_2 nanoparticles achieved pronounced activity and stability in photocatalytic hydrogen generation. The solar H_2 generation rate obtained was around 10 mmol·hour^{-1}·g^{-1}, which is about two orders of magnitude greater than the yields of most semiconduc-tor photocatalysts [22]. The energy conversion efficiency for solar hydrogen production, defined as the ratio of the energy of solar produced hydrogen to the energy of incident light, reached 24% for surface-disordered black TiO_2 nanoparticles [22].

Li and coworkers also developed hydrogen-treated metal oxides such as TiO_2 [11], ZnO [24], WO_3[25], and $BiVO_4$ for PEC and photocatalytic applications. Figure 3.1a shows the rate of photogeneration of H_2 collected for air-annealed ZnO and hydrogen-treated ZnO nanorod arrays and ZnO nanorod powder [24]. Hydrogen-treated ZnO shows significantly higher hydrogen production rate than pristine ZnO, indicating hydrogen treatment is effective in enhancing ZnO photoactivity. Importantly, the hydrogen-treated ZnO exhibits excellent photostability in photogeneration of hydrogen, without obvious decay in the production rate within 24 hours (Fig. 3.1b). Moreover, it is noteworthy that there was no visible light photoactivity observed for hydrogen-treated ZnO, although the color of ZnO changed from white to black after hydrogen treatment [24]. The enhanced photocatalytic

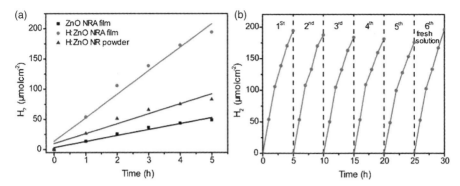

FIGURE 3.1   (a) Photocatalytic hydrogen generation rate collected for ZnO nanorod arrays (NRA) film, hydrogen-treated ZnO (H:ZnO) NRA film, and H:ZnO nanorod (NR) powder in a solution containing 0.1 M Na_2SO_3 and 0.1 M Na_2S under white light irradiation. (b) Cycling performance of H:ZnO NRA films. *Source*: Reproduced with permission from Lu et al. [24]. (See color insert.)

performance of ZnO for water splitting was attributed to the increased electrical conductivity as a result of introduction of oxygen vacancies, which is a shallow donor for ZnO.

Despite the various approaches that have been demonstrated to be effective in improving the optical and electronic properties of metal oxides such as ZnO and TiO_2, the large bandgap of these metal oxides is still one of the major factors limiting their performance. Therefore, it is highly desirable to develop small bandgap semiconductors for photocatalytic hydrogen generation.

3.2.3   Metal Oxynitrides/Metal Nitrides/Metal Phosphides

Metal nitrides or metal oxynitrides such as Ta_3N_5 and TaON have attracted increasing attention due to the relatively smaller bandgap energies compared with their oxide counterparts [15, 26]. It is known that the valence band of metal oxides is dominated by O2p orbital, and the conduction band edges consist predominantly of the empty orbitals of metal ions. The low lying O 2p orbital imposes an intrinsic limitation on the electronic band structure for simultaneously achieving favorable bandgap energy and band edge positions. In comparison with metal oxides, metal nitrides and metal oxynitrides have more negative valence band edge due to the hybridization of N 2p with O 2p orbitals [27]. For example, Ta_2O_5 has a bandgap of 4.2 eV, while Ta_3N_5 and TaON have narrower bandgap energies of 2.1 and 2.5 eV, respectively [28]. TaON and Ta_3N_5 can be prepared by nitridation of Ta_2O_5 at elevated temperature in ammonia gas [28]. The UV–visible diffuse reflectance spectra of Ta_2O_5 and TaON show that the absorption edge of TaON was around 500 nm with an estimated bandgap of 2.5 eV, which was red shifted by about 170 nm from that of Ta_2O_5 after nitridation. The time course of H_2 evolution on TaON under irradiation of visible light ($\lambda > 420$ nm) showed that H_2 production was preceded continuously without obvious decrease in activity [28].

Furthermore, metal phosphides such as GaP also have been studied for photocatalytic hydrogen generation. Compared with Ga_2O_3 (4.8 eV), GaP has a smaller bandgap of 2.79 eV (direct energy gap) and 2.3 eV (indirect energy gap), and suitable band edge positions for water reduction. Recently, Sun et al. reported a surfactant free, self-seeded solution–liquid–solid (SLS) approach for large-scale synthesis of high quality colloidal GaP nanowires [18]. These high quality GaP nanowires with clean surface hold great potentials as visible light photocatalyst for water splitting. Transmission electron microscopy (TEM) images proved that the as-prepared GaP nanowires are uniform single crystals with a length of 1–2 μm and diameter of 30 nm (Fig. 3.2a and Fig. 3.2b). Importantly, the prepared GaP nanowires can be assembled into a large nanowire membrane by filtration (Fig. 3.2c). The capability of using GaP nanowires as photocatalyst to drive hydrogen evolution from water reduction under visible light illumination with methanol as hole scavenger was demonstrated. To improve hydrogen evolution, GaP nanowires were further decorated with Pt nanoparticles. As shown in Figure 3.2d, the hydrogen evolution rate of Pt-modified GaP (with a low Pt loading of 2 wt%) nanowire was an order of magnitude higher than that of GaP nanowire. The linear profiles suggest the hydrogen generation rate is constant and GaP nanowires are stable for photocatalytic hydrogen generation within 12 hours (Fig. 3.2d).

Metal oxynitrides, nitrides, and phosphides have been demonstrated to be useful as visible light photocatalysts for hydrogen generation. However, the long-term stability is still a concern for these materials due to the self-oxidation during photocatalytic reactions.

3.2.4   Metal Chalcogenides

CdS and CdSe nanoparticles are two common metal chalcogenides used as photocatalysts for hydrogen generation [9,29]. They have suitable band edge positions for water splitting, and their bandgap energies are tunable via controlled variation of particle size. For example, Li et al. reported CdS cluster-modified graphene nanosheets for visible light-driven photocatalytic hydrogen production [30]. The layered graphene as a supporting matrix can

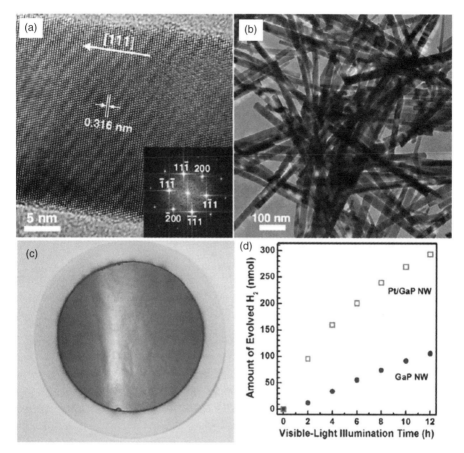

FIGURE 3.2   (a–b) TEM images of GaP nanowires. The inset in (a) shows the indexed FFT pattern of the image, indicating the wire is a single crystal with a growth axis of [111] direction. (c) Photograph of a large GaP nanowire membrane on a PVDF filter membrane. *Source*: Reproduced with permission from Sun et al. [18]. (See color insert.)

efficiently suppress charge recombination by facilitating charge transfer of the photoexcited carriers. The CdS-graphene composite was prepared using a solvothermal method in which graphene oxide and cadmium acetate are the precursors of graphene and CdSe, respectively [30]. TEM images showed that the graphene sheets were decorated with many small particles (Fig. 3.3a), and these particles are single crystals with an average size of around 3 nm [30]. The CdS-graphene composite was further modified with Pt catalyst for hydrogen evolution. Figure 3.3c illustrates the three possible mechanisms of electron transfer on the composite under visible light irradiation: (1) to Pt nanoparticles deposited on the surface of CdS cluster; (2) to

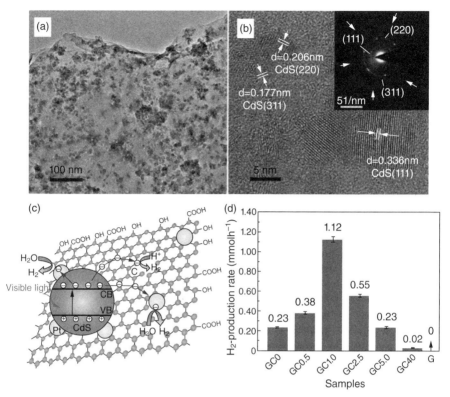

FIGURE 3.3 (a,b) TEM images of graphene sheet decorated with CdS clusters. Inset: SAED pattern collected at the composite structure. (c) Schematic illustration of the charge separation and transfer in the graphene-CdS system under visible light. (d) Comparison of the visible light photocatalytic activity of graphene–CdS systems with different graphene loading for the H_2 production using 10 vol% lactic acid aqueous solution as a sacrificial reagent and 0.5 wt% Pt as a co-catalyst. *Source*: Reproduced with permission from Li et al. [30]. (See color insert.)

graphene sheet; and (3) to Pt nanoparticles located on the graphene sheets. Eventually, the electrons will react with the absorbed proton to form H_2 [30]. The H_2 production rate was noticeably increased with the loading of even a small amount of graphene (0.5–2.5 wt%). When 1.0 wt% of graphene was loaded, the H_2 production rate reached the optimal value of 1.12 mmol h^{-1}, with a quantum efficiency of 22.5% at 420 nm. The enhanced photoactivity of CdS-graphene composite was attributed to the efficient charge separation of photoexcited carriers. On the other hand, the overloading of black graphene led to shielding of the active sites on the active sites on the catalyst surface and rapidly reduced the intensity of light through the depth of the reaction solution and thus the photoactivity. The results also prove that the bare graphene sheets without CdS cluster were not active for photocatalytic

H_2 production under the same measurement conditions [30]. This work demonstrated the potential of graphene as a support for CdS nanoparticles for photocatalytic hydrogen generation.

Recently, Han et al. recently reported a robust and highly active system of Ni catalyst-modified CdSe quantum dots for solar hydrogen generation [29]. The photoexcited electrons on CdSe nanocrystals are shuttled to Ni catalyst and reduce protons to hydrogen while the photoexcited holes are used to oxidize ascorbic acid [29]. The hydrogen evolution performances of CdSe nanocrystals with and without Ni-catalyst modification were measured under light illumination. The linear profiles indicate that the hydrogen generation rate is constant for over 360 hours, suggesting high stability over a long period of time. In comparison with the system without Ni catalyst, the Ni-modified CdSe (denoted as Ni-CdSe) nanocrystals exhibited more than one order of magnitude enhancement in hydrogen generation rate. The results suggested that the Ni catalyst facilitates the electron transfer for proton reduction and further boosts the photoactivity of CdSe. The Ni-CdSe system achieved an optimal turnover number over 600,000 mol of H_2 per mole of catalyst, which is a benchmark value for photocatalytic H_2 generation among different photocatalysts such as metal oxides, metal nitrides, and metal chalocogenides. The quantum yield obtained is around 36% at the wavelength of 520 nm with ascorbic acid as hole sacrificial reagent. Importantly, the homogenous nickel catalyst is from just common inexpensive nickel salts, such as nickel nitrate, nickel chloride, and nickel acetate [29].

3.2.5 Conclusion

Meal oxides, oxynitrides, and chalcogenides have been extensively explored for photocatalytic hydrogen generation. The recent breakthroughs in developing catalyst-modified nanostructured photocatalysts suggested the feasibility of large-scale production of solar hydrogen. Nevertheless, there are still several outstanding questions that need to be addressed. For instance, although these photocatalysts have suitable electronic band structure for water splitting, most of these photocatalytic systems require the addition of sacrificial reagents, due to large overpotential for water oxidation and their instability in oxidative environment. Moreover, spontaneous water reduction requires the conduction band of photocatalyst more negative than water reduction potential. Semiconductors such as α-Fe_2O_3 has a favorable bandgap for visible light absorption but relatively positive conduction potential that limits its application as photocatalyst for hydrogen generation. Alternatively, these semiconductors can be used as electrode materials for PEC hydrogen generation, by applying an external bias, as discussed in the next section.

3.3   PHOTOELECTROCHEMICAL METHODS

3.3.1   Background

Solar-driven photoelectrochemical (PEC) water splitting represents a sustainable and environmentally friendly method to produce hydrogen. A PEC cell consists of at least one semiconductor photoelectrode, which can harvest solar light [31]. n- and p-type semiconductors are preferred for the photoanode and photocathode, respectively. Under light irradiation with photon energy equal to or exceeding the bandgap energy of the semiconductor photoelectrode, the electrons are excited from the valence band to the unoccupied conduction band. The band bending at the semiconductor–electrolyte interface or an applied bias will facilitate the separation of photogenerated electrons and holes. The electrons will transfer to the cathode–electrolyte interface to reduce protons to generate hydrogen ($2H^+ + 2e^- \rightarrow H_2$), while the holes will oxidize water molecules to produce oxygen ($H_2O + 2h^+ \rightarrow 2H^+ + \frac{1}{2} O_2$) at the anode–electrolyte interface. Depending on the band edge positions of electrode materials, additional external potential may be required to achieve water splitting.

The development of high performance photoelectrodes (anode and cathode) represents a major challenge for PEC water splitting. A good photoelectrode should have favorable bandgap for efficient visible light absorption and band edge positions that straddle the redox potentials of water electrolysis. Moreover, it must be highly resistant to photocorrosion, electrochemically stable in aqueous solution in reductive (cathode) and oxidative (anode) environment, and have good electrical conductivity. And finally it must be inexpensive. In the meantime, for PEC devices with semiconductor anode and cathode, it is also critical that the photocurrent of anode and cathode are matched and the device can be operated at zero external bias as shown in Figure 3.4 [31].

3.3.2   Photocathode for Water Reduction

Cuprous oxide, Cu_2O, is an attractive p-type metal oxide for PEC hydrogen production with a direct bandgap of 2 eV, for which a theoretical photocurrent density of 14.7 mA cm^{-2} and a solar to hydrogen conversion efficiency of 18% are predicted [32]. Cu_2O also possesses favorable energy band positions, with the conduction band lying 0.7 V negative of the hydrogen evolution potential and the valence band lying just positive of the oxygen evolution potential. However, the redox potentials for the reduction and oxidation of Cu_2O lie within its bandgap, resulting in its poor electrochemical stability. Cu(I) is more prone to be reduced to Cu metal, rather than

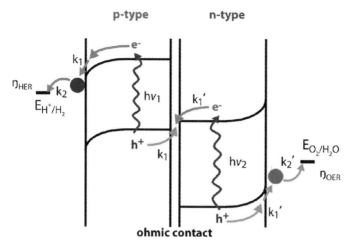

FIGURE 3.4   Schematic energy diagram of PEC water splitting with (a) photoanode, (b) photocathode, and (c) n-type photoanode and p-type photocathode. *Source*: Reproduced with permission from Liu et al. [31]. (See color insert.)

reducing water. Recently, the photo-instability of Cu_2O in water was addressed by deposition of a protective layer on the electrode surface by atomic layer deposition (ALD) method (Fig. 3.5a) [32]. The protective oxide layer forms a staggered type II band offset with Cu_2O, so photogenerated electrons can flow from the Cu_2O conduction band through the protective layer to the electrolyte for water reduction. Additionally, the n-type oxide layer should have a conduction band negative to the hydrogen evolution potential, without reductive degradation reaction at the potentials within the bandgap. Moreover, the oxide should have low overpotential for water reduction. The Cu_2O films used for this study were synthesized by electrodeposition method, with a thickness of 1.3 µm [32]. Individual Cu_2O grains of the film were about 1 µm in size with a cubic morphology, and had a predominant (111) orientation (Fig. 3.5b). Various metal oxide coatings were tested and Cu_2O can be stabilized by coating with 5 × (4 nm ZnO/0.17 nm Al_2O_3)/11 nm TiO_2 and Pt nanoparticles (Fig. 3.5a). Under AM 1.5 illumination, the as-deposited bare Cu_2O produced a photocurrent of −2.4 mA cm^{-2} at 0.25 V versus the reversible hydrogen electrode (RHE) in a nitrogen purged 1 M Na_2SO_4 electrolyte buffered at a pH of 4.9 (Fig. 3.5c). However, the cathodic current quickly decreased to zero, indicating that bare Cu_2O is not stable for PEC water reduction (Fig. 3.5c, inset). On the contrary, the surface-protected Cu_2O with 5 × (4 nm ZnO/0.17 nm Al_2O_3)/11 nm TiO_2/Pt shows substantially enhanced photoactivity and

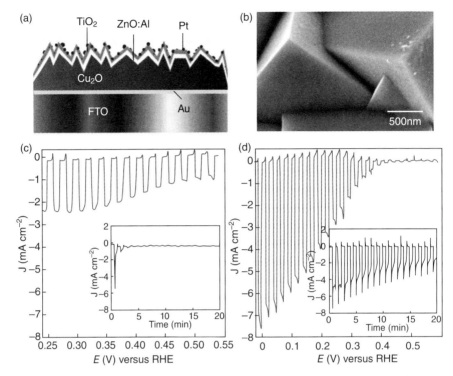

FIGURE 3.5   (a) Schematic presentation of the electrode structure. (b) Scanning electron micrograph showing a top view of the electrode after ALD of $5 \times (4$ nm $ZnO/0.17$ nm $Al_2O_3)/11$ nm TiO_2 followed by electrodeposition of Pt nanoparticles. (c) Current-potential characteristics in 1 M Na_2SO_4 solution under chopped AM 1.5 light illumination for the bare Cu_2O electrode, (d) for the as-deposited $5 \times (4$ nm $ZnO/0.17$ nm $Al_2O_3)/11$ nm TiO_2. The insets show respective photocurrent transient for the electrodes held at 0 V versus RHE in chopped light illumination with N_2 purging. *Source*: Reproduced with permission from Paracchino et al. [32]. (See color insert.)

photostability for water reduction (Fig. 3.5d). A photocurrent of 7 mA cm^{-2} was obtained at 0.25 V versus RHE for the protected Cu_2O electrode. Importantly, 78% of photocurrent retention was achieved on the protected Cu_2O after 20-min illumination (Fig. 3.5d, inset). The Faradaic efficiency of H_2 generation is close to 100%, indicating the photocurrent decay was not due to the degradation of the photoactive materials. The decay was attributed to the presence of Ti^{3+} traps in the TiO_2 layer. Since the Fermi level of TiO_2 in the dark is close to the water reduction potential, the electrons were not readily injected into the electrolyte and accumulated in the protective layer as long-lived Ti^{3+} states. This study demonstrated an effective strategy to stabilize Cu_2O by coating protective metal oxide layers.

Silicon is the second most abundant element in earth's crust, and has been extensively used in electronic and photovoltaic devices. Photoelectrochemical H_2 generation at Si/electrolyte interfaces has also been studied for decades [33–35]. The large surface overpotential for hydrogen evolution is the major limitation for silicon-based photocathode. Enormous research efforts have been placed to modify the silicon surface with electrocatalysts such as Pt to suppress the overpotential [36]. In order to improve the photoactivity of silicon for water reduction, Oh et al. reported to use p-type silicon nanowire array photocathode fabricated via metal-catalyzed electroless etching (Fig. 3.6a). The silicon nanowire arrays were further impregnated with Pt nanoparticles (Fig. 3.6b). The nanowire-arrayed electrode significantly

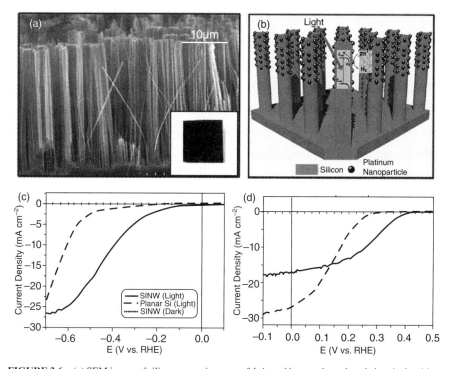

FIGURE 3.6   (a) SEM image of silicon nanowire arrays fabricated by metal-catalyzed chemical etching; inset is the photograph of ~10 mm × 10 mm silicon nanowire array sample with low reflection. (b) Schematic of silicon nanowire arrayed photoelectrode. Photon absorbed by silicon nanowire generates minority carrier, which drifts to semiconductor/electrolyte interface where H^+ is reduced to H_2. Silicon nanowires are impregnated with Pt nanoparticles that serve as electrocatalysts for water reduction. (c) PEC performance of bare silicon nanowire and planar silicon film. (d) PEC performance of Pt modified planar silicon and nanowire silicon. *Source*: Reproduced with permission from Oh et al. [36]. (See color insert.)

increased the semiconductor/electrolyte interfacial area, which can improve the kinetics of H_2 generation. In addition, the silicon nanowire arrays could function as an antireflective or light trapping layer to minimize reflection of incident light, and therefore enhancing the light absorption. As shown in Figure 3.6c, the onset potential on the silicon nanowire photocathode was 0.2 V more positive than that of planar Si, confirming the large surface area could suppress the surface over-potential of hydrogen generation. In the presence of Pt electrocatalyst, photocurrent onset potential was further positively shifted to 0.4 V, which is the lowest onset potential ever reported for *p*-type silicon photocathode [36].

To date, the report of *p*-type semiconductor photoelectrodes for water splitting is still rare, compared with *n*-type semiconductors. In addition to Cu_2O and silicon discussed earlier, there are several other *p*-type materials reported for water splitting, including GaP [37], $GaInP_2$ [38], and InP [39] Electrochemical stability and surface kinetics are still the two major issues for these semiconductor electrodes. The deposition of protective layer and electrochemical catalysts are two of the best strategies to address these limitations. On the other hand, in order to achieve nonbiased PEC device for water splitting, it is equally important to develop high performance photoanode with photocurrent matching with the photocathode.

3.3.3 Photoanode for Water Oxidation

Various *n*-type semiconductors, such as silicon [40], metal oxides [11, 13, 25], and metal nitrides [15, 26], have been explored for use as photoanodes. Among them, metal oxides are of special interest due to their simple synthetic process, excellent chemical stability during water oxidation, and suitable band edge positions [4]. These metal oxides can be divided into binary metal oxides, such as TiO_2, ZnO, Fe_2O_3, and WO_3, and ternary metal oxides, such as $BiVO_4$, $SrTiO_3$, and $CuWO_4$. In this section, we will review the recent studies of metal oxide photoanodes for PEC water oxidation.

TiO_2 and α-Fe_2O_3 are binary metal oxides commonly used for PEC water oxidation. TiO_2 has suitable band edge positions for water splitting; however, its large bandgap energy of 3.0–3.2 eV limits visible light absorption [11, 13]. In contrast, α-Fe_2O_3 has a favorable bandgap of 2.1–2.2 eV with substantial visible light absorption, but its conduction band is more positive to the hydrogen evolution potential. Additionally, α-Fe_2O_3 has poor electrical conductivity and short carrier diffusion length, which leads to significant charge recombination loss [41, 42].

Recently, Wang et al. have developed a general strategy to incorporate oxygen vacancies into metal oxides, such as TiO_2, ZnO, and WO_3 via

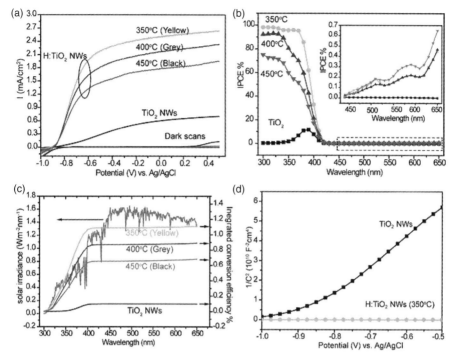

FIGURE 3.7 (a) Linear sweep voltammograms collected on pristine TiO_2 nanowire and hydrogen-treated TiO_2 (H:TiO_2) nanowires annealed at temperature of 350, 400, and 450°C. (b) IPCE spectra of pristine TiO_2 and H:TiO_2 nanowires. The inset is the magnified IPCE spectra that highlighted in the dashed box. (c) Simulated solar-to-hydrogen efficiencies for the pristine TiO_2 and H:TiO_2 samples as a function of wavelength, by integrating their IPCE spectra collected at −0.6 V versus Ag/AgCl with a standard AM 1.5G solar spectrum. (d) Mott–Schottky plots collected at a frequency of 5 kHz in the dark for pristine TiO_2 and H:TiO_2 nanowire. *Source*: Reproduced with permission from Wang et al. [11]. (See color insert.)

hydrogen treatment at elevated temperatures. Oxygen vacancies serve as shallow donors that can improve the electrical conductivity of metal oxides [11, 25]. Notably, hydrogen-treated samples showed substantially increased photocurrent density, compared with pristine TiO_2 (Fig. 3.7a). A maximum photocurrent density of around 2.5 mA cm^{-2} was obtained for the hydrogen-treated TiO_2 at 0 V versus Ag/AgCl in 1.0 M NaOH aqueous solution [11]. Incident photon-to-current efficiency (IPCE) analysis suggested that the enhanced photocurrent was due to improved photoactivity of TiO_2 in the UV region (Fig. 3.7b). The increased IPCE values were attributed to enhanced charge collection efficiency as expected for hydrogen-treated TiO_2 that has improved electrical conductivity. By integrating the IPCE spectra with standard AM 1.5G spectrum, a simulated maximum solar to hydrogen conversion efficiency of 1.1% was obtained for hydrogen-treated TiO_2 (Fig. 3.7c).

Electrochemical impendence studies proved that the donor density of TiO_2 was substantially increased after hydrogen treatment (Fig. 3.7d). More importantly, hydrogenation has been demonstrated to be a general strategy for increasing donor density of metal oxides. Likewise, enhanced donor density and improved photocurrent density have been observed in other metal oxides, including WO_3 and ZnO [24,25].

Visible light absorption of wide bandgap metal oxides can be enhanced by chemical doping using elements such as nitrogen and carbon [8, 21, 43]. For example, Park et al. reported carbon-doped, vertically aligned TiO_2 nanotube arrays for PEC water splitting by annealing TiO_2 nanotube in the mixed gas of CO and CO_2 [43]. The carbon-doped TiO_2 showed substantially increased photocurrent density under visible light illumination (>420 nm) than pristine TiO_2 nanotubes. Similarly, Hoang et al. reported nitrogen-doped TiO_2 nanowire arrays for PEC water splitting under visible light [8]. A substantial amount of nitrogen (up to 1.08 atomic %) can be incorporated into the TiO_2 lattice via nitridation in ammonia gas flow at a relative low temperature. After nitrogen doping, the white-colored pristine TiO_2 became yellow. The absorption spectrum confirmed visible light absorption of TiO_2. IPCE analysis shows that nitrogen-doped TiO_2 exhibits visible light photoactivity up to 500 nm. Hoang et al. also combined hydrogen treatment and nitrogen doping of TiO_2 nanowire arrays and found a synergistic effect between oxygen vacancies and nitrogen dopant in the PEC performance [21]. The hydrogen-treated and nitrogen-doped TiO_2 showed the best performance with visible light photoactivity, compared with TiO_2 samples treated with hydrogen and nitrogen alone.

As mentioned earlier, hematite (α-Fe_2O_3) exhibits poor PEC performance for water splitting, due to its poor electrical conductivity and short charge diffusion length. Element doping using elements such as Si [44], Ti [41], and Sn [45] have been explored to increase the donor density and electrical conductivity of hematite. Recently, Ling et al. found that hematite can be activated via the incorporation of oxygen vacancies, which function as shallow donors in hematite [46]. The formation of oxygen vacancy was achieved by annealing β-FeOOH nanowire arrays in an oxygen-deficient condition (mixture of nitrogen and air), whereas Fe^{3+} can be reduced to Fe^{2+}. X-ray photoelectron spectroscopy (XPS) Fe2p confirmed the existence of Fe^{2+} in the hematite after the treatment (Fig. 3.8a). Electrochemical impedance studies supported the electrical conductivity and donor density being significantly enhanced, which was attributed to the incorporation of oxygen vacancies (Fig. 3.8b). As a result, the photocurrent density of hematite was increased by orders of magnitude to more than 3 mA cm^{-2} at 1.4 V versus NHE. The substantial enhancement was believed to be due to the improved

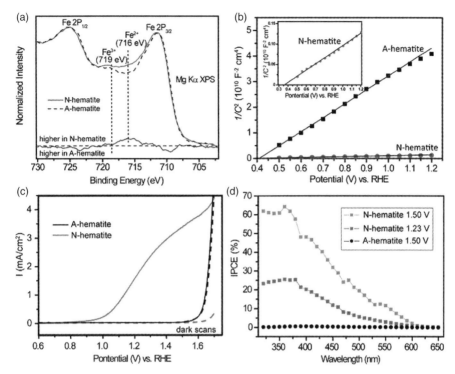

FIGURE 3.8 (a) Overlay of Fe 2p XPS spectra of air annealed hematite (denoted as: A-hematite) and oxygen-deficient hematite (denoted as: N-hematite), together with their different spectrums. The dashed lines highlight the satellite peaks of Fe^{2+} and Fe^{3+}. (b) Mott–Schottky plots measured for A-hematite and N-hematite. Inset: magnified Mott-schottky plot of N-hematite. (c) Linear sweep voltammograms collected on A-hematite and N-hematite under a simulated solar light of 100 mW cm^{-2} and dark condition with a scan rate of 10 mV s^{-1}. (d) The corresponding IPCE spectra for A-hematite and N-hematite collected at potentials of 1.23 and 1.5 V versus RHE. *Source*: Reproduced with permission from Ling et al. [46]. (See color insert.)

charge transport in the hematite. Moreover, IPCE studies also showed enhanced photoactivity in the entire wavelength region that is consistent with the bandgap of hematite.

Previous studies have primarily focused on binary metal oxides for PEC water splitting. However, one of the fundamental constrains of these binary metal oxides is that their valence bands are typically composed of O2p character, which lie far more positive than the water oxidation potential. As a result, some energy will be wasted during water oxidation and limit the solar conversion efficiency. Recently, ternary metal oxides have attracted renew attentions on PEC water splitting, as their electronic bands are formed by atomic orbitals from more than one element and the modulation of the stoichiometric ratio of the elements could be used to fine-tune the potentials of valence and conduction bands as well as the bandgap energy. Therefore, ternary metal oxides show promise for developing high efficient photoelectrode for water splitting with suitable bandgap and band edge positions.

As an example, $BiVO_4$ is a direct bandgap ternary metal oxide semiconductor with a favorable bandgap of 2.3–2.5 eV for solar light absorption. Furthermore, its conduction band is close to 0V versus RHE at pH = 0, as a result of the hybridization of empty Bi 6p orbitals with antibonding V 3d-O 2p states, which reduce the need for external bias for PEC water splitting. However, charge transport and interfacial charge transfer have been found to be key limiting factors for its PEC performance. A number of methods have been explored to address these limitations. Element doping with Mo and W has been found to be effective in increasing the PEC performance of $BiVO_4$, as M and W introduce shallow donors that improve the separation and transport of photoexcited carriers. In addition, porous electrode was fabricated to provide a large surface area and shorten the diffusion distance for minority carriers. For instance, Pilli et al. reported cobalt phosphate (Co–Pi)-modified, Mo-doped $BiVO_4$ photoelectrode for solar water oxidation [47]. Mo-doped $BiVO_4$ was prepared using a surfactant assisted metal organic decomposition technique at 500°C. The Mo-doped $BiVO_4$ film exhibited absorption in the visible region up to 520 nm. The bandgap was estimated to be around 2.4 eV. Co–Pi catalyst was deposited on the surface of Mo-$BiVO_4$ by electrochemical deposition method. The role of Co–Pi catalyst is used to reduce the overpotential of $BiVO_4$ for water oxidation. Importantly, the photocurrent density of Mo-doped $BiVO_4$ electrode was enhanced compared with that of the $BiVO_4$ electrode. The photocurrent of porous $BiVO_4$ is higher than nonporous $BiVO_4$. Moreover, significantly enhanced photocurrents were observed for the Co–Pi catalyzed electrodes in the entire potential range, compared with unmodified electrodes. The Co Pi electrodeposited electrode also exhibited around 150 mV cathodic shift from the onset potential for PEC water oxidation, as compared with unmodified $BiVO_4$.

Although $BiVO_4$ has favorable bandgap, its relatively poor charge transport properties causes a significant electron–hole recombination loss. Recently, Hong et al. reported heterojunction $BiVO_4/WO_3$ electrodes that shown enhanced photoactivity for water oxidation [48]. $BiVO_4$ and WO_3 form a type II junction at the interface, and photogenerated electrons from the conduction band of $BiVO_4$ are thermodynamically favorable for transferring into the WO_3 layer. Meanwhile, the photoexcited holes in WO_3 are also favorable for transferring to $BiVO_4$ for water oxidation. The formation of heterojunction facilitates charge separation and thus suppresses the electron–hole recombination. The heterojunciton electrodes were fabricated by layer-by-layer deposition of WO_3 and $BiVO_4$ on a conducting glass. The PEC studies showed that the photocurrent onsets of WO_3 and $BiVO_4/WO_3$ were 0.25 and 0.05 V versus Ag/AgCl, indicating the flat-band potential negatively shifted with $BiVO_4$ coating. The $BiVO_4/WO_3$ composite electrodes also show improved photocurrent densities compared to pristine WO_3. The results supported the notation that heterojunctions could improve the PEC water oxidation of $BiVO_4$.

Other ternary metal oxides, such as $CuWO_4$ (2.25 eV), $InVO_4$ (2.0 eV), and $FeVO_4$ (2.7 eV), have also been synthesized and studied as photocatalysts for water oxidation and pollutant degradation. Studies reported to date have shown that ternary metal oxides hold promise as high performance photo-electrode for solar water splitting.

3.3.4 Conclusion

With control over size, morphology, and element doping of semiconductors, their performance for PEC hydrogen generation have been significantly improved. While significant progress has been made, several challenging issues still remain to be addressed for solar hydrogen generation. For example, while element doping has been proved to be effective in enhancing the PEC performance of metal oxides, the long-term stability of element-doped metal oxides is a concern for practical applications. Also, there is still not a single photoelectrode that can achieve highly efficient water splitting under nonbiased condition. Although 12 % solar-to-hydrogen conversion efficiency has been achieved in hybrid devices that integrated PEC cell with solar cell, the cost and the electrochemical instability of the devices are limitations for large scale production. To date, the solar to hydrogen conversion efficiency of PEC systems are still low, especially compared with the conversion efficiency that can be achieved in a solar cell-powered electrolysis process. It is highly desirable and challenging to develop new low cost and high performance photoelectrodes for solar hydrogen generation.

3.4 SUMMARY

A number of semiconductor materials have been tested for photocatalytic and photoelectrochemical hydrogen generation and have made significant progress in enhancing the solar conversion efficiency. There are outstanding challenges that need to be addressed. For example, most of the photocatalytic systems require the addition of hole sacrificial reagents such as methanol or sulfide ions, which is a major drawback for photocatalytic hydrogen generation. Development of high efficiency hydrogen and oxygen evolution reaction catalysts is high desirable. For PEC water splitting, there is still not a single photoelectrode can achieve highly efficient water splitting under non-biased condition. Nonbiased PEC water splitting can be achieved by coupling semiconductor photoanode and photocathode. The charge balance between these two electrodes is a major challenge. The combination of theoretical calculation and experimental efforts to design and develop better photocatalytic and photoelectrochemical materials is fundamentally important for solar hydrogen generation.

REFERENCES

1. Mazloomi, K., Sulaiman, N.B., Moayedi, H. Electrical efficiency of electrolytic hydrogen production. *International Journal of Electrochemical Science* **2012**, 7, 3314–3326.

2. Djafour, A., Matoug, M., Bouras, H., Bouchekima, B., Aida, M.S., Azoui, B. Photovoltaic-assisted alkaline water electrolysis: Basic principles. *International Journal of Hydrogen Energy* **2011**, *36*, 4117–4124.

3. Tachibana, Y., Vayssieres, L., Durrant, J. R. Artificial photosynthesis for solar water-splitting. *Nature Photonics* **2012**, *6*, 511–518.

4. Wang, G.M., Ling, Y.C., Li, Y. Oxygen-deficient metal oxide nanostructures for photoelectrochemical water oxidation and other applications. *Nanoscale,* **2012**, *4*, 6682–6691.

5. Kudo, A., Miseki, Y. Heterogeneous photocatalyst materials for water splitting. *Chemical Society Reviews* **2009**, *38*(1), 253–278.

6. Fujishima, A., Honda, K. Electrochemical photolysis of water at a semiconductor electrode. *Nature* **1972**, *238*, 37–38.

7. Li, Y., Zhang, J.Z. Hydrogen generation from photoelectrochemical water splitting based on nanomaterials. *Laser & Photonics Reviews* **2010**, *4*, 517–528.

8. Hoang, S., Guo, S.W., Hahn, N.T., Bard, A.J., Mullins, C.B. Visible light driven photoelectrochemical water oxidation on nitrogen-modified TiO$_2$ nanowires. *Nano Letters* **2012**, *12*, 26–32.

9. Zhu, H.M., Song, N.H., Lv, H.J., Hill, C.L., Lian, T.Q. Near unity quantum yield of light-driven redox mediator reduction and efficient H$_2$ generation using colloidal nanorod heterostructures. *Journal of the American Chemical Society* **2012**, *134*, 11701–11708.

10. Hwang, Y.J., Boukai, A., Yang, P.D. High density n-Si/n-TiO$_2$ core/shell nanowire arrays with enhanced photoactivity. *Nano Letters* **2009**, *9*, 410–415.

11. Wang, G.M., Wang, H.Y., Ling, Y.C., Tang, Y.C., Yang, X.Y., Fitzmorris, R.C., Wang, C.C., Zhang, Z.J., Li, Y. Hydrogen-treated TiO$_2$ nanowire arrays for photoelectrochemical water splitting. *Nano Letters* **2011**, *11*, 3026–3033.

12. Wang, G.M., Yang, X.Y., Qian, F., Zhang, J.Z., Li, Y. Double-sided CdS and CdSe quantum dot co-sensitized ZnO nanowire arrays for photoelectrochemical hydrogen generation. *Nano Letters* **2010**, *10*, 1088–1092.

13. Yang, X.Y., Wolcott, A., Wang, G.M., Sobo, A., Fitzmorris, R.C., Qian, F., Zhang, J.Z., Li, Y. Nitrogen-doped ZnO nanowire arrays for photoelectrochemical water splitting. *Nano Letters* **2009**, *9*, (6), 2331–2336.

14. Zhang, J., Bang, J.H., Tang, C.C., Kamat, P.V. Tailored TiO_2-$SrTiO_3$ heterostructure nanotube arrays for improved photoelectrochemical performance. *ACS Nano* **2010**, *4*, 387–395.

15. Banerjee, S., Mohapatra, S.K., Misra, M. Synthesis of TaON nanotube arrays by sono-electrochemical anodization followed by nitridation: A novel catalyst for photoelectro-chemical hydrogen generation from water. *Chemical Communications* **2009**, *46*, 7137–7139.

16. Ito, S., Thampi, K.R., Comte, P., Liska, P., Gratzel, M. Highly active meso-microporous TaON photocatalyst driven by visible light. *Chemical Communications* **2009**, *2*, 268–270.

17. Ma, S.S.K., Hisatomi, T., Maeda, K., Moriya, Y., Domen, K. Enhanced water oxidation on Ta_3N_5 photocatalysts by modification with alkaline metal salts. *Journal of the American Chemical Society* **2012**, *134*, 19993–19996.

18. Sun, J.W., Liu, C., Yang, P.D. Surfactant-free, large-scale, solution-liquid-solid growth of gallium phosphide nanowires and their use for visible-light-driven hydrogen production from water reduction. *Journal of the American Chemical Society* **2011**, *133*, 19306–19309.

19. Zhang, R.Q., Liu, X.M., Wen, Z., Jiang, Q. Prediction of silicon nanowires as photocata-lysts for water splitting: band structures calculated using density functional theory. *Journal of Physical Chemistry C* **2011**, *115*, 3425–3428.

20. Wang, F.Y., Yang, Q.D., Xu, G., Lei, N.Y., Tsang, Y.K., Wong, N.B., Ho, J.C. Highly active and enhanced photocatalytic silicon nanowire arrays. *Nanoscale* **2011**, *3*, 3269–3276.

21. Hoang, S., Berglund, S.P., Hahn, N.T., Bard, A.J., Mullins, C.B. Enhancing visible light photo-oxidation of water with TiO_2 nanowire arrays via cotreatment with H_2 and NH_3: synergistic effects between Ti^{3+} and N. *Journal of the American Chemical Society* **2012**, *134*, 3659–3662.

22. Chen, X., Liu, L., Yu, P.Y., Mao, S.S. Increasing solar absorption for photocatalysis with black hydrogenated titanium dioxide nanocrystals. *Science* **2011**, *331*, 746–750.

23. Thompson, T.L., Yates, J.T. Surface science studies of the photoactivation of TiO_2-new photochemical processes. *Chemical Reviews* **2006**, *106*, 4428–4453.

24. Lu, X.H., Wang, G.M., Xie, S.L., Shi, J.Y., Li, W., Tong, Y.X., Li, Y. Efficient photocata-lytic hydrogen evolution over hydrogenated ZnO nanorod arrays. *Chemical Communications* **2012**, *48*, 7717–7719.

25. Wang, G.M., Ling, Y.C., Wang, H.Y., Yang, X.Y., Wang, C.C., Zhang, J.Z., Li, Y. Hydrogen-treated WO_3 nanoflakes show enhanced photostability. *Energy & Environmental Science* **2012**, *5*, 6180–6187.

26. Tabata, M., Maeda, K., Higashi, M., Lu, D.L., Takata, T., Abe, R., Domen, K. Modified Ta3N5 powder as a photocatalyst for O-2 Evolution in a two-step water splitting system with an iodate/iodide shuttle redox mediator under visible light. *Langmuir* **2010**, *26*, 9161–9165.

27. Fang, C. M., Orhan, E., de Wijs, G.A., Hintzen, H.T., de Groot, R.A., Marchand, R., Saillard, J.Y., de With, G. The electronic structure of tantalum (oxy)nitrides TaON and Ta$_3$N$_5$. *Journal of Materials Chemistry* **2011**, *11*, 1248–1252.

28. Hitoki, G., Takata, T., Kondo, J.N., Hara, M., Kobayashi, H., Domen, K. An oxynitride, TaON, as an efficient water oxidation photocatalyst under visible light irradiation (lambda $\Leftarrow$ 500 nm). *Chemical Communications* **2002**, *16*, 1698–1699.

29. Han, Z.J., Qiu, F., Eisenberg, R., Holland, P.L., Krauss, T.D. Robust photogeneration of H2 in water using semicondutor nanocrystals and a nickel catalyst. *Science* **2012**, *338*, 1321–1324.

30. Li, Q., Guo, B.D., Yu, J. G., Ran, J.R., Zhang, B.H., Yan, H.J., Gong, J.R. Highly efficient visible-light-driven photocatalytic hydrogen production of CdS-cluster-decorated graphene nanosheets. *Journal of The American Chemical Society* **2011**, *133*, 10878–10884.

31. Liu, C., Dasgupta, N.P., Yang, P.D. Semiconductor nanowires for artificial photosynthesis. *Chemistry of Materials* **2013**, in press (DOI: 10.1021/cm4023198).

32. Paracchino, A., Laporte, V., Sivula, K., Gratzel, M., Thimsen, E. Highly active oxide photocathode for photoelectrochemical water reduction. *Nature Materials* **2010**, *10*, 456–461.

33. Bookbinder, D.C., Lewis, N.S., Bradley, M.G., Bocarsly, A.B., Wrighton, M.S. Photoelectrochemical reduction of N, N-dimethyl-4,4-Bipyridinium in aqueous media at p type silicon-sustained photogeneration of a species capable of evolving hydrogen. *Journal of The American Chemical Society* **1979**, *101*, 7721–7723.

34. Bookbinder, D.C., Bruce, J.A., Dominey, R.N., Lewis, N.S., Wrighton, M.S. Synthesis and characterization of a photosensitive interface for hydrogen generation- chemically modified p-type semiconducting silicon photocathode. *Proceedings of the National Academy of Sciences of the United States of America-Physical Sciences* **1980**, *77*, 6280–6284.

35. Boettcher, S.W., Spurgeon, J.M., Putnam, M.C., Warren, E.L., Turner-Evans, D.B., Kelzenberg, M.D., Maiolo, J.R., Atwater, H.A., Lewis, N.S. Energy-conversion properties of vapor-liquid-solid-grown silicon wire-array photocathodes. *Science* **2010**, *327*, 185–187.

36. Oh, I., Kye, J., Hwang, S. Enhanced photoelectrochemical hydrogen production from silicon nanowire array photocathode. *Nano Letters* **2012**, *12*, 298–302.

37. Liu, C., Sun, J.W., Tang, J.Y., Yang, P.D. Zn-doped p-type gallium phosphide nanowire photocathodes from a surfactant-free solution synthesis. *Nano Letters* **2012**, *12*, 5407–5411.

38. Khaselev, O., Turner, J.A. Electrochemical stability of p-GaInP$_2$ in aqueous electrolytes toward photoelectrochemical water splitting. *Journal of the Electrochemical Society* **1998**, *145*, 3335–3339.

39. Lee, M.H., Takei, K., Zhang, J.J., Kapadia, R., Zheng, M., Chen, Y.Z., Nah, J., Matthews, T. S., Chueh, Y.L., Ager, J.W., Javey, A. p-Type InP nanopillar photocathodes for efficient solar-driven hydrogen production. *Angewandte Chemie-International Edition* **2012**, *51*, 10760–10764.

40. Pijpers, J.J.H., Winkler, M.T., Surendranath, Y., Buonassisi, T., Nocera, D.G. Light-induced water oxidation at silicon electrodes functionalized with a cobalt oxygen-evolving catalyst. *Proceedings of the National Academy of Sciences of the United States of America* **2012**, *108*, 10056–10061.

41. Wang, G.M., Ling, Y. C., Wheeler, D.A., George, K.E.N., Horsley, K., Heske, C., Zhang, J.Z., Li, Y. Facile synthesis of highly photoactive α-Fe_2O_3-based films for water oxidation. *Nano Letters* **2011**, *11*, 3503–3509.

42. Wheeler, D.A., Wang, G.M., Ling, Y.C., Li, Y., Zhang, J.Z. Nanostructured hematite: synthesis, characterization, charge carrier dynamics, and photoelectrochemical properties. *Energy & Environmental Science* **2012**, *5*, 6682–6702.

43. Park, J. H., Kim, S., Bard, A. J., Novel carbon-doped TiO_2 nanotube arrays with high aspect ratios for efficient solar water splitting. *Nano Letters* **2006**, *6*, 24–28.

44. Cesar, I., Sivula, K., Kay, A. Zboril, R., Grätzel, M. Influence of feature size, film thickness, and silicon doping on the performance of nanostructured hematite photoanodes for solar water splitting. *Journal of Physical Chemistry C* **2009**, *113*, 772–782.

45. Ling, Y.C., Wang, G.M., Wheeler, D.A., Zhang, J.Z., Li, Y. Sn-doped hematite nanostructures for photoelectrochemical water splitting. *Nano Letters* **2011**, *11*, 2119–2125.

46. Ling, Y., Wang, G., Reddy, J., Wang, C., Zhang, J., Li, Y. Influence of oxygen content on thermal activation of hematite nanowires. *Angewandte Chemie International Edition* **2012**, *51*, 4074–4079.

47. Pilli, S.K., Furtak, T.E., Brown, L.D., Deutsch, T.G., Turner, J.A., Herring, A.M. Cobalt-phosphate (Co-Pi) catalyst modified Mo-doped $BiVO_4$ photoelectrodes for solar water oxidation. *Energy & Environmental Science* **2011**, *4*, 5028–5034.

48. Hong, S.J., Lee, S., Jang, J.S., Lee, J.S. Heterojunction $BiVO_4$/WO_3 electrodes for enhanced photoactivity of water oxidation. *Energy & Environmental Science* **2011**, *4*, 1781–1787.

4

Biohydrogen Generation and Other Methods

4.1 BASICS ABOUT BIOHYDROGEN

An alternative for hydrogen generation is through biological processes, so-called biohydrogen generation. Biohydrogen produced from biorenewables is a promising alternative for a sustainable energy source. Biohydrogen is a renewable biofuel produced from biorenewable feedstocks by chemical, thermochemical, biological, biochemical, and biophotolytical methods [1]. Biohydrogen is particularly attractive for rural areas of the world since it can potentially be carried out at low cost and with more primitive techniques. In addition, there are several reasons for biohydrogen to be considered as a relevant technology by both industrialized and developing countries. First, it is less energy-intensive compared with other processes for hydrogen generation, such as steam reforming of natural gas, coal gasification, or electrolysis of water. Second, less undesired by-products, such as CO_2, are produced in the process compared with some of the other techniques. This is because using biomass instead of fossil fuels to produce hydrogen reduces the net amount of CO_2 released to the atmosphere, as the CO_2 released when the biomass is gasified was previously absorbed from the atmosphere and fixed by photosynthesis in the growing plants [2].

Currently, there are two major approaches for using biological organisms for hydrogen production. The first is direct production through

Hydrogen Generation, Storage, and Utilization, First Edition. Jin Zhong Zhang, Jinghong Li, Yat Li, and Yiping Zhao.
© 2014 John Wiley & Sons, Inc. Published 2014 by John Wiley & Sons, Inc.

"photobiohydrogen" with microorganisms capable of using solar photons to separate oxygen from water. The second is production using biotechnologies-based microorganisms that produce hydrogen naturally.

However, obtaining hydrogen from biomass has major challenges in practice. Currently, there are no completed technology platforms for large-scale biohydrogen generation even though small or laboratory scale demonstrations have been done. The yield of hydrogen is low from biomass since the hydrogen content in biomass is low to begin with (approximately 6% versus 25% for methane) and the energy content is low due to the 40% oxygen content of biomass. Due to the potential advantages mentioned earlier, substantial research efforts have been made to explore the potential of biohydrogen generation.

4.2 PATHWAYS OF BIOHYDROGEN PRODUCTION FROM BIOMASS

The methods available for hydrogen production from biomass can be divided into two main categories: thermochemical and biological routes. The major biomass-to-hydrogen pathways are shown in Figure 4.1 [3]. Hydrogen can be produced from biorenewable feedstock via thermo chemical conversion processes such as pyrolysis, gasification, steam gasification, steam reforming of bio-oils, and supercritical-water gasification. Biological production of hydrogen can be classified into the following groups: (i) biophotolysis of water using green algae and blue-green algae (cyanobacteria), (ii) photofermentation, (iii) dark fermentation, and (iv) hybrid reactor system.

The advantage of the thermochemical process is that its overall efficiency (thermal to hydrogen) is higher ($\eta \sim 52\%$) and production cost is lower [4]. The yield of hydrogen that can be produced from biomass is relatively low, 16–18% based on dry biomass weight [5]. Hydrogen yields and energy contents, compared with biomass energy contents obtained from processes with biomass, are shown in Table 4.1 [6]. In the pyrolysis and gasification processes, water gas shift is used to convert the reformed gas into hydrogen, and pressure swing adsorption is used to purify the product. Compared with other biomass thermo chemical gasification such as air gasification or steam gasification, the supercritical water gasification can directly deal with the wet biomass without drying, and has high gasification efficiency in lower temperature [7]. The major disadvantage of these processes is that the decomposition of the biomass feedstock leads to char and tar formation [8]. In order to optimize the process for hydrogen production, efforts have been made by researchers to test hydrogen production from biomass gasification/pyrolysis with various biomass types and under different operating conditions.

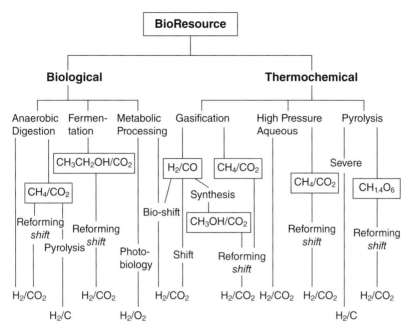

FIGURE 4.1 Pathways from biomass-to-hydrogen. *Source*: Reproduced with permission from Milne [3].

TABLE 4.1   Comparison of Hydrogen Yields Are Obtained by Use of Three Different Processes

Processes	Hydrogen Yield (wt%)	Hydrogen Energy Contents/ Biomass Energy Content
Pyrolysis + catalytic reforming	12.6	91
Gasification + shift reaction	11.5	83
Biomass + steam + except heat (theoretical maximum)	17.1	124

Source: Reproduced with permission from Wang et al. [6].

An example of oil palm shell compared with physic nut waste is listed in Table 4.2 [9].

Biological hydrogen production processes are found to be more environmentally friendly and less energy demanding compared with thermochemical and electrochemical processes [10]. Researchers have started to investigate hydrogen production with anaerobic bacteria since 1980s [11]. Biological production of hydrogen (biohydrogen) as a byproduct of microorganism

TABLE 4.2   Hydrogen Production from Conversion of Oil Palm Shell and Physic Nut Waste

Type Biomass/Temperature	Gas Production (vol%)
Oil palm shell	
773 K	3.56
973 K	12.58
1173 K	33.49
Physic nut	
773 K	8.22
973 K	9.29
1173 K	11.63

Source: Reproduced with permission from Sricharoenchaikul et al. [9].

metabolism is an exciting new area of technology development that offers the potential production of usable hydrogen from a variety of renewable sources [12]. There are three types of microorganisms of biohydrogen generation: cyano-bacteria, anaerobic bacteria, and fermentative bacteria. A promising method is the biological production of hydrogen by fermentation. The production of hydrogen from biomass by fermentation is one of the routes that can contribute to a future sustainable hydrogen economy. The amount of hydrogen produced from glucose is affected by fermentation pathways and liquid end products [13]. Butyric acid and acetic acid constitute more than 80% of total end products [14]. Theoretically, 4 mol of H_2 can be produced from 1 mol of glucose in acetate-type fermentation, however only 2 mol of H_2 are produced when butyrate is the main fermentation product. To date, many studies have been done on fermentative hydrogen production from pure sugars and from feedstock, such as byproducts from the agricultural and food industry, municipal waste, or wastewaters [15]. Anaerobic digestion of solid organic waste, including municipal and agricultural wastes and wastewater sludge, is one such renewable source for H_2 production. However, continual H_2 production using this process has limitations, one of which is the low yields of hydrogen currently produced from the fermentation of even the simplest sugars [16]. A combination of dark and photofermentation in a two-stage hybrid system has been found to improve the overall yield of hydrogen [17]. Anaerobic bacteria decompose glucose or starch via acetate fermentative metabolism as the first step, and photosynthetic bacteria convert the resultant acetate to hydrogen in another reactor as the second stage. The hydrogen yield is increased by twofolds in comparison to that using only dark fermentation [18].

4.3   THERMOCHEMICAL CONVERSION OF BIOMASS TO HYDROGEN

4.3.1   Hydrogen from Biomass via Pyrolysis

Pyrolysis of biomass is a promising route for the production of solid (char), liquid (tar and other organics), and gaseous products as possible alternative sources of energy. The most interesting temperature range for the production of the pyrolysis products is between 625 and 775 K [19, 20]. Depending on the operating conditions, the pyrolysis process can be divided into three subclasses: conventional (slow) pyrolysis, fast pyrolysis, and flash pyrolysis. Slow pyrolysis of biomass is associated with high charcoal continent, but the fast pyrolysis is associated with tar at low temperature (675–775 K) and/or gas at high temperature [21]. At present, the preferred technology is fast or flash pyrolysis at high temperatures with very short residence time. Table 4.3 indicates the product distribution obtained from different processes of pyrolysis process [22]. Although most pyrolysis processes are designed for biofuel production, hydrogen can be produced directly through fast or flash pyrolysis if high temperature and sufficient volatile phase residence time are allowed as follows [23]:

$$\text{Biomass} + \text{Heat} \rightarrow H_2 + CO + CH_4 + \text{Other products.} \qquad (4.1)$$

Methane and other hydrocarbon vapors can be converted into hydrogen and carbon monoxide (CO) by steam reforming:

$$CH_4 + H_2O \rightarrow CO + 3H_2. \qquad (4.2)$$

TABLE 4.3   Product Distribution Obtained from Different Processes of Pyrolysis Process

Thermal Degradation	Residence Time (s)	Upper Temperature (K)	Product Yield (%)		
			Char	Liquid	Gas
Slow pyrolysis	200	600	32–38	28–32	25–29
	120	700	29–33	30–35	32–36
	90	750	26–32	27–34	33–37
	60	850	24–30	26–32	35–43
	30	950	22–28	23–29	40–48
Fast pyrolysis	5	700	22–27	53–59	12–16
	4	750	17–23	58–64	13–18
	3	800	14–19	65–72	14–20
	2	850	11–17	68–76	15–21
	1	950	9–13	64–71	17–24
Gasification	1500	1250	8–12	4–7	81–88

Source: Reproduced with permission from Balat [22].

A water–gas shift reaction can be applied in order to increase the hydrogen production:

$$CO + H_2O \rightarrow CO_2 + H_2. \tag{4.3}$$

Pyrolysis of biomass is a complex function of the experimental conditions, under which the pyrolysis process proceeds. The important factors, which affect the yield and composition of the volatile fraction liberated, are biomass species, chemical and structural composition of biomass, particle size, temperature, heating rate, residence time, atmosphere, pressure and reactor configuration [24]. Yield of products resulting from biomass pyrolysis include charcoal, from a low temperature and low heating rate process, liquid products, from a low temperature, high heating rate, and short gas residence time process, and fuel gas, from a high temperature, low heating rate, and long gas residence time process.

The yields of hydrogen-rich gases via pyrolysis were related to the temperature. Increasing the pyrolysis temperature resulted in an increase in the hydrogen yield as a percentage of the total gases evolved [25]. The percentage of hydrogen in gaseous products by pyrolysis from the samples of hazelnut shell, tea waste, and spruce wood increased from 36.8% to 43.5%, 41.0% to 53.9%, and 40.0% to 51.5% by volume, respectively, when the final pyrolysis temperature was increased from 700 to 950 K. One of the methods to increase the hydrogen yield is to apply catalytic pyrolysis. Three types of biomass feedstock, olive husk, cotton cocoon shell and tea waste were pyrolyzed at about 775–1025 K in the presence of $ZnCl_2$ and catalyst-to-biomass ratios of 6.5–17 by weight [26]. The highest yield of hydrogen-rich gas (70.3%) was achieved from olive husk using about 13% $ZnCl_2$ at about 1025 K. The K_2CO_3 and Na_2CO_3 as catalysts also affected on yield of gaseous products from various biomass species with pyrolysis. The effect of K_2CO_3 and Na_2CO_3 on pyrolysis depends on the biomass species. The catalytic effect of Na_2CO_3 was greater than that of K_2CO_3 for the cotton cocoon shell and tea factory waste, but the catalytic effect of K_2CO_3 was greater for the olive husk [26]. The yields of hydrogen-rich gas from pyrolysis of agricultural residues in the presence of Na_2CO_3 were different at different temperatures. Results shown in Figure 4.2 indicate that the highest yield of hydrogen-rich gas was obtained from the walnut shell sample [27]. Among the different metal oxides, Al_2O_3 and Cr_2O_3 exhibit better catalytic effect than the others [23]. The use of noble metals (Rh, Ru, and Pt) in large-scale industrial steam reforming is not common because of their relative high cost [28].

Hydrogen can also be produced by catalytic steam reforming of bio-oil or its fractions [29, 30]. Hydrogen production from bio-oil provides a new route

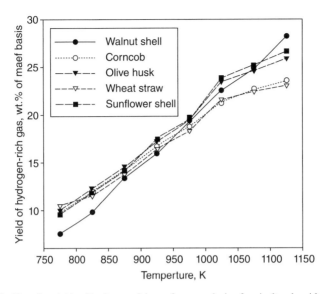

FIGURE 4.2 Plots for yields of hydrogen-rich gas from pyrolysis of agricultural residues versus temperature in the presence of 30% Na_2CO_3. *Source*: Reproduced with permission from Demirbas [27].

for the utilization of bio-oil. Hydrogen production from renewable bio-oil is an attractive idea for fuel, energy, and agricultural applications. In recent years, hydrogen production via steam reforming of bio-oil has attracted more and more attention. But because of the complicated composition of bio-oil and carbon deposition on catalyst surface in the reaction process, currently, studies have mainly focused on steam reforming of model compounds in bio-oil and reforming catalysts [31]. The bio-oil can be stored and shipped to a centralized facility where it is converted to hydrogen via catalytic steam reforming and shift conversion [32]. Catalytic steam reforming of bio-oil at 1025–1125 K over a Ni-based catalyst is a two-step process that includes the shift reaction [33]:

$$\text{Bio-oil} + H_2O \rightarrow CO + H_2 \qquad (4.4)$$

$$CO + H_2O \rightarrow CO_2 + H_2. \qquad (4.5)$$

The overall stoichiometry gives a maximum yield of 0.172 g $H_2 \cdot g^{-1}$ bio-oil (11.2% based on wood) [33].

$$CH_{1.9}O_{0.7} + 1.26\ H_2O \rightarrow CO_2 + 2.21\ H_2. \qquad (4.6)$$

In reality, this yield will always be lower because both the steam reforming and water–gas shift reactions are reversible, resulting in the presence of some

CO and CH_4 in the product gas. In addition, thermal cracking that occurs parallel to reforming produces carbonaceous deposits [34].

4.3.2 Hydrogen from Biomass via Gasification

Gasification of biomass has been identified as a possible system for producing renewable hydrogen, which is beneficial to exploiting biomass resources and developing a highly efficient clean way for large-scale hydrogen production, with less dependence on insecure fossil energy sources [35]. In general, the gasification temperature is higher than that of pyrolysis and the yield of hydrogen from the gasification is higher than that of the pyrolysis.

Biomass gasification can be considered as a form of pyrolysis, which takes place at higher temperatures and produces a mixture of gases with H_2 content ranging 6–6.5% [36]. The synthetic gas produced by the gasification of biomass is made up of H_2, CO, CH_4, N_2, CO_2, and O_2; and tar is also formed that is often removed with a physical dust removal method [37]. The product distribution and gas composition depend on many factors, including the gasification temperature and the reactor type. The most important gasifier types are fixed bed (updraft or downdraft fixed beds), fluidized bed, and entrained flow gasifiers. All these gasifiers need to include significant gas conditioning along with the removal of tar and inorganic impurities and the subsequent conversion of CO to H_2 by water gas shift reaction, as discussed in the pyrolysis section. Table 4.4 shows typical gas composition data as obtained from commercial wood and charcoal downdraft gasifiers operated on low to medium moisture content fuels [38].

Gasification technologies provide the opportunity to convert renewable biomass feedstocks into clean fuel gases or synthesis gases. The synthesis gas includes mainly hydrogen and carbon monoxide (H_2 + CO), which is also called bio-syngas [39, 40]. Table 4-5 shows the composition of bio-syngas from biomass gasification [36]. Hydrogen production is the largest use of syngas. Biomass can be converted to bio-syngas by non-catalytic, catalytic, and steam gasification processes.

TABLE 4.4 Typical Gas Composition Data as Obtained from Commercial Wood and Charcoal Downdraft Gasifiers Operated on Low to Medium Moisture Content Fuels (Wood 20%, Charcoal 7%)

Component	H_2 (%)	CO_2 (%)	CH_4 (%)	CO (%)	N_2 (%)	Heating Value (MJ·m^{-3})
Wood gas	12–20	9–15	2–3	17–22	50–54	5–5.9
Charcoal gas	4–10	1–3	0–2	28–32	55–65	4.5–5.6

Source: Reproduced with permission from Stassen and Knoef [38].

TABLE 4.5   Composition of Bio-Syngas from Biomass Gasification

Constituents	% by Volume (Dry and Nitrogen Free)
Carbon monoxide (CO)	28–36
Hydrogen (H_2)	22–32
Carbon dioxide (CO_2)	21–30
Methane (CH_4)	8–11
Ethene (C_2H_4)	2–4
Benzene–toluene–xylene (BTX)	0.84–0.96
Ethane (C_2H_5)	0.16–0.22
Tar	0.15–0.24
Others (NH_3, H_2S, HCl, dust, ash, etc.)	<0.021

Reproduced with permission from Demirbras [36].

Steam gasification is a promising technology for thermochemical hydrogen production from biomass. Hydrogen is produced from the steam gasification of legume straw and pine sawdust [41], hazelnut shell [42], paper, yellow pine woodchips [43], mosses, algae [19], wood sawdust [44], wheat straw [45], and waste wood [46].

Steam reforming C1–C5 hydrocarbons, nafta, gas oils, and simple aromatics are commercially practiced, well-known processes. Steam reforming of hydrocarbons, partial oxidation of heavy oil residues, selected steam reforming of aromatic compounds, and gasification of coals and solid wastes to yield a mixture of H_2 and CO, followed by water-gas shift conversion to produce H_2 and CO_2, are well established processes [47]. Steam reforming and so-called dry or CO_2 reforming occur according to the following reactions and are usually promoted by the use of catalysts:

$$C_nH_m + n\,H_2O \leftrightarrow CO + (n + m/2)\,H_2 \qquad (4.7)$$

$$C_nH_m + n\,CO_2 \leftrightarrow (2n)\,CO + (m/2)\,H_2. \qquad (4.8)$$

Modeling of biomass steam gasification to synthesis gas is a challenge because of the variability (composition, structure, reactivity, and other physical properties) of the raw material and because of the severe conditions (temperature, residence time, and heating rate) required [48]. The yield of H_2 from steam gasification increases with increasing water-to-sample (W/S) ratio [49] The yields of H_2 from steam gasification increase with increasing temperature. The yield of hydrogen-rich gaseous product in the gaseous products from the black liquor steam gasification run (W/S = 1.9) increased from 38.0% to 50.3% with increasing temperature from 975 to 1325 K [50].

The effect of catalyst on gasification products is very important. The use of catalysts did not affect the gas yields, but the composition of the gases was strongly influenced. The content of H_2 and CO_2 increased, while that of CO decreased. A drastic reduction in the content of organic compounds could also be observed. Because the char yields remained almost constant compared to an equivalent no catalytic thermal run, the increase in the content of hydrogen was probably due to the influence of the catalyst on the water gas shift reaction. Dolomite, Ni-based catalysts and alkaline metal oxides are widely used as gasification catalysts [22] The yields of hydrogen from biomass with the use of dolomite in the fluidized-bed gasifier and the use of nickel-based catalysts in the fixed bed reactor downstream from the gasifier were investigated by Lv et al. [51]. They obtained a maximum hydrogen yield (130.28 g $H_2 \cdot kg^{-1}$ biomass) over the temperature range of 925–1125 K. K_2CO_3 catalyst shows a destructive effect on the organic compounds, and H_2 and CO_2 form at the end of the catalytic steam reforming process [52]. The catalytic steam gasification of biomass in a lab-scale fixed bed reactor was carried out in order to evaluate the effects of particle size at different bed temperatures on the gasification performance [53]. With decreasing particle size, the dry gas yield, carbon conversion efficiency and H_2 yield increased, and the content of char and tar decreased.

4.3.3 Hydrogen from Biomass via Supercritical Water (Fluid–Gas) Extraction

The supercritical fluid extraction (SFE) is a separation technology that uses supercritical fluid solvent. Fluids cannot be liquefied above the critical temperature, regardless of the pressure applied, but may reach the density close to the liquid state. Every fluid is characterized by a critical point, which is defined in terms of the critical temperature and critical pressure. Water is a supercritical fluid above 647.2 K and 22.1 MPa [54, 55].

Supercritical water (SCW) possesses properties very different from those of liquid water. The dielectric constant of SCW is much lower, and the number of hydrogen bonds is much lower and their strength is weaker. As a result, high temperature water behaves like many organic solvents so that organic compounds have complete miscibility with SCW. Moreover, gases are also miscible in SCW, thus a SCW reaction environment provides an opportunity to conduct chemistry in a single fluid phase that would otherwise occur in a multiphase system under conventional conditions [56].

The biomass gasification in SCW is a complex process, but the overall chemical conversion can be represented by the simplified net reaction:

$$CH_xO_y + (2-y)\,H_2O \rightarrow CO_2 + (2-y+x/2)\,H_2, \tag{4.9}$$

where x and y are the elemental molar ratios of H/C and O/C in biomass, respectively. The reaction product is syngas, whose quality depends on x and y. The reaction (Eq. 4.9) is an endothermic. It is known from the reaction (Eq. 4.9) that water is not only the solvent but also a reactant, and the hydrogen in the water is released by the gasification reaction [57].

Compared with other biomass thermochemical gasifications, such as air gasification or steam gasification, the SCW gasification has high gasification efficiency at lower temperature and can deal directly with wet biomass without drying [58]. Hydrogen production by biomass gasification in SCW is a promising technology for utilizing high moisture content biomass. Another advantage of SCW reforming is that the H_2 is produced at a high pressure, which can be stored directly, thus avoiding the large energy expenditures associated with its compression. The cost of hydrogen production from SCW gasification of wet biomass was several times higher than the current price of hydrogen from steam methane reforming. Biomass is gasified in supercritical water at a series of temperatures and pressures during different resident times to form a product gas composed of H_2, CO_2, CO, CH_4, and a small amount of C_2H_4 and C_2H_6 [36, 59]. SCW is a promising reforming media for the direct production of hydrogen at 875–1075 K temperatures with a short reaction time (2–6 seconds). As the temperature is increased from 875 to 1075 K, the H_2 yield increases from 53% to 73% by volume, respectively [59, 60]. Only a small amount of hydrogen is formed at low temperatures, indicating that direct reformation reaction of ethanol as a model compound in SCW is favored at high temperatures (>975 K) [59]. With an increase in the temperature, the hydrogen and carbon dioxide yields increase, while the methane yield decreases.

4.3.4   Comparison of Thermochemical Processes

In general, the gasification temperature is higher than that of pyrolysis and the yield of hydrogen from the gasification is higher than that of the pyrolysis. The yield of hydrogen from conventional pyrolysis of corncob increases from 33% to 40% with increasing of temperature from 775 to 1025 K. The yields of hydrogen from steam gasification increase from 29% to 45% for (W/S) = 1 and from 29% to 47% for (W/S) = 2 with increasing of temperature from 975 to 1225 K [45]. Hydrogen yields and energy contents, compared with biomass energy contents obtained from processes with biomass, are shown in Table 4.6 [61].

Demirbas investigated the yield of hydrogen from supercritical fluid extraction (SFE), pyrolysis, and steam gasification of wheat straw and olive waste at different temperatures [45]. The highest yields (% dry and ash free

TABLE 4.6   Comparison of Hydrogen Yields Are Obtained by Use of Three Different Processes

Processes	Hydrogen Yield (wt%)	Hydrogen Energy Contents/ Biomass Energy Content
Pyrolysis + catalytic reforming	12.6	91
Gasification + shift reaction	11.5	83
Biomass + steam + except heat (theoretical maximum)	17.1	124

Source: Reproduced with permission from Wang et al. [61].

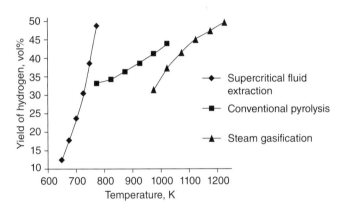

FIGURE 4.3   Plots for yield of hydrogen from supercritical fluid extraction, pyrolysis, and steam gasification [(W/S) = 2] of beech wood at different temperatures. *Source*: Reproduced with permission from Demirbas [45].

basis) were obtained from the pyrolysis (46%) and steam gasification (55%) of wheat straw, while the lowest yields were from olive waste. He also investigated the yield of hydrogen from SFE, pyrolysis and steam gasification of beech wood at different temperatures. Distilled water was used in the SFE (the critical temperature of pure water is 647.7 K). Results of this study are shown in Figure 4.3. From Figure 4.3, the yield of hydrogen from SFE was considerably high (49%) at lower temperatures. The pyrolysis was carried out at the moderate temperatures and steam gasification at the highest temperatures.

4.4   BIOLOGICAL PROCESS FOR HYDROGEN PRODUCTION

Hydrogen produced from water, renewable organic wastes or biomass, either biologically (biophotolysis and fermentation) or photobiologically (photode-

TABLE 4.7   The Main Advantages of Different Biological Hydrogen Production Processes

Process	Advantages
Direct biophotolysis	It can produce H_2 directly from water and sunlight. Solar conversion energy increased by 10-fold as compared with trees and crops
Indirect biophotolysis	It can produce H_2 from water. Has the ability to fix N_2 from atmosphere
Photofermentation	A wide spectral light energy can be used by these bacteria. It can use different waste materials, such as distillery effluents, and waste
Dark fermentation	It can produce H_2 all day long without light. A variety of carbon sources can be used as substrates. It produces valuable metabolites such as butyric, lactic, and acetic acids as by products. It is anaerobic process, so there is no O_2 limitation problem.
Hybrid reactor system (combined dark and photofermentation)	Two-stage fermentation can improve the overall yield of hydrogen

Source: Reproduced with permission from Nath and Das [66].

composition), is termed "biohydrogen." Biological hydrogen production processes are found to be more environment friendly and less energy intensive as compared with thermochemical and electrochemical processes [62]. Researchers have started to investigate hydrogen production with anaerobic bacteria since 1980s [63–65]. The main advantages of different biological hydrogen production processes are given in Table 4.7 [66].

The processes of biological hydrogen production can be broadly classified into two distinct groups. One is light-dependent and the other is light-independent process. Specific ways in which microorganisms can produce H_2 include biophotolysis of water using green algae and blue-green algae (cyanobacteria), photofermentation, dark fermentation, and hybrid reactor system [67].

There are three types of microorganisms of biohydrogen generation: cyanobacteria, anaerobic bacteria, and fermentative bacteria. The cyanobacteria directly decompose water to biohydrogen and oxygen in the presence of light energy by photosynthesis. Photosynthetic bacteria use organic substrates like organic acids. Anaerobic bacteria use organic substances as the sole source of electrons and energy, converting them into biohydrogen. Biological hydrogen can be generated from plants by biophotolysis of water using microalgae (green algae and cyanobacteria), fermentation of organic compounds, and photodecomposition of organic compounds by photo-synthetic bacteria [63]. All processes of biological hydrogen production are fundamentally

TABLE 4.8   Classification of Hydrogenases

Classification	Occurrence/Source	Structure	Features	
			Localization	Function
Ni–Fe	Anaerobic, photosynthetic bacteria, cyanobacteria	Heterodimeric, multimeric	Membrane-bound, cytoplasmic, periplasmic	Uptake of hydrogen
Ni–Fe–Se	Sulphate-reducing bacteria, methanogenes	Oligomeric	Membrane-bound, cytoplasmic	Oxidation of hydrogen
Fe	Photo-synthetic bacteria, anaerobic fermentative bacteria, cyanobacteria, green algae, protozoan	Monomeric, heteromeric	Cytoplasmic, mambrane-bound, periplasmic chloroplast, hydrogenosomes	Production of hydrogen
Metal-free	Methanogens	Monomeric	Cytoplasmic	Formation of hydrogen

Source: Reproduced with permission from Das et al. [68].

dependent upon the presence of a hydrogen-producing enzyme. Hydrogenases and nitrogenases are the known enzymes which catalyze biological hydrogen production. Hydrogenases are the key enzymes for the biological hydrogen production, which can be classified as uptake hydrogenases and reversible hydrogenases. Uptake hydrogenases, such as Ni–Fe hydrogenases and Ni–Fe–Se hydrogenases, act as important catalysts for hydrogen consumption as follows [23]:

$$H_2 \rightarrow 2e^- + 2H^+ \tag{4.10}$$

Reversible hydrogenases, as indicated by its name, have the ability to produce H_2 as well as consume H_2 depending on the reaction condition,

$$H_2 \leftrightarrow 2e^- + 2H^+. \tag{4.11}$$

Despite increasingly conspicuous diversity in many respects, hydrogenase can be classified broadly into three distinct classes: Ni–Fe hydrogenase, Fe-hydrogenase, and metal-free hydrogenase [68]. Classification of hydrogenases is given in Table 4.8.

4.4.1   Biophotolysis of Water Using Microalgae

Biophotolysis is the action of light on biological systems that results in the dissociation of a substrate usually water into molecular hydrogen and oxygen. Photosynthetic bacteria (e.g., Rhodobactor) can use broad organic substrates, including lactic and acetic acids, as the energy and carbon source under light irradiation. Photoautotrophic green algae and cyanobacteria use sunlight and carbon dioxide as the sole sources for energy and carbon [1]. Based on a preliminary engineering and economic analysis, biophotolysis processes must achieve close to an overall 10% solar energy conversion efficiency to be competitive with alternatives sources of renewable hydrogen [69].

4.4.1.1 Direct Biophotolysis

Direct biophotolysis of hydrogen production is a biological process that utilizes light energy and photosynthetic systems of microalgae to convert water into chemical energy.

$$2\ H_2O + Light\ energy \rightarrow 2\ H_2 + O_2. \tag{4.12}$$

The solar energy is absorbed by the pigments at photosystem I (PSI), or photosystem II (PSII) or both, which raises the energy level of electrons from water oxidation when they are transferred from PSI via PSII to ferredoxin. The concept of "direct biophotolysis" envisions light-driven simultaneous O_2 evolution on the oxidizing side of PSII and H_2 production on the reducing side of PSI, with a maximum $H_2 : O_2$ (mol : mol) ratio of $2 : 1$ [70].

Since hydrogenase is sensitive to oxygen, it is necessary to maintain the oxygen content at a low level under 0.1% so that hydrogen production can be sustained. This condition can be obtained by the use of green algae Chlamydomonas reinhardtii that can deplete oxygen during oxidative respiration [23, 67]. Reported hydrogen production rates using this method are approximately $0.07\ mmol \cdot h^{-1}$ per liter [71, 72].

4.4.1.2 Indirect Biophotolysis

In indirect biophotolysis, the problem of sensitivity of the hydrogen evolving process to oxygen is usually circumvented by separating oxygen and hydrogen [72–74]. The concept of indirect biophotolysis involves the following four steps: [23] (1) biomass production by photosynthesis, (2) biomass concentration, (3) aerobic dark fermentation yielding 4 mol hydrogen/mol glucose in the algae cell, along with 2 mol of acetates, and (4) conversion of 2 mol of acetates into hydrogen. In a typical indirect biophotolysis, cyanobacteria can synthesize and evolve hydrogen through photosynthesis by following reactions:

$$12\ H_2O + 6\ CO_2 + Light\ energy \rightarrow C_6H_{12}O_6 + 6\ O_2 \tag{4.13}$$

$$C_6H_{12}O_6 + 12\ H_2O + Light\ energy \rightarrow 12\ H_2 + 6\ CO_2. \tag{4.14}$$

Hydrogen production by cyanobacteria has been studied for over three decades and has revealed that efficient photoconversion of H_2O to H_2 is influenced by many factors. Rates of H_2 production by nonnitrogen-fixing Cyanobacteria range from 0.02 μmol $H_2 \cdot mg^{-1}$ chl a/h (*Synechococcus* PCC 6307) to 0.40 μmol $H_2 \cdot mg^{-1}$ chl a/h (*Aphanocapsa montana*) [75].

4.4.2 Photofermentation

Purple nonsulfur bacteria evolve molecular hydrogen catalyzed by nitrogenase under nitrogen-deficient conditions using light energy and reduced compounds (organic acids) [76]. These bacteria themselves are not powerful enough to split water. However, under anaerobic conditions, these bacteria are able to use simple organic acids, like acetic acid, or even hydrogen disulfide as electron donor. These electrons are transported to the nitrogenase by ferredoxin using energy in the form of adenosine triphosphate (ATP). When nitrogen is not present, this nitrogenase enzyme can reduce proton into hydrogen gas again using extra energy in the form of ATP [72]. The reaction can be given as

$$C_6H_{12}O_6 + 12\ H_2O + \text{Light energy} \rightarrow 12\ H_2 + 6\ CO_2. \qquad (4.15)$$

Among the various bioprocesses capable of hydrogen production, photofermentation is favored due to relatively higher substrate-to-hydrogen yields and, its ability to trap energy under a wide range of the light spectrum and versatility in sources of metabolic substrates with promise for waste stabilization [77]. In photofermentation processes, the yield of the order of 80% has been achieved [72]. However, these processes have three main drawbacks [23]: (1) use of nitrogenase enzyme with high-energy demand, (2) low solar energy conversion efficiency, and (3) demand for elaborate anaerobic photobioreactors covering large areas.

4.4.3 Dark Fermentation

Hydrogen can be produced by anaerobic bacteria, grown in the dark on carbohydrate-rich substrates. Dark fermentation of carbohydrate-rich substrates as biomass presents a promising route of biological hydrogen production, compared with photosynthetic routes. Anaerobic hydrogen fermenting bacteria can produce hydrogen continuously without the need for photoenergy. Dark hydrogen fermentation can be performed at different temperatures: mesophilic (298–313 K), thermophilic (313–338 K), extreme-thermophilic (338–353 K), or hyperthermophilic (>353 K). While direct and indirect photolysis systems produce pure H_2, dark fermentation processes produce a mixed biogas containing primarily H_2 and CO_2, but which may also contain lesser amounts of methane (CH_4), CO, and/or hydrogen sulfide (H_2S) [75]. Glucose yields different amount of hydrogen depending on the fermentation pathways and liquid end products. A maximum of 4 mol hydrogen is theoretically produced from 1 mol of glucose with acetic acid as the end product, while a maximum of 2 mol hydrogen is theoretically produced from 1 mol of glucose with butyrate as the end product [78]:

$$C_6H_{12}O_6 + 2\ H_2O \rightarrow 2\ CH_3COOH + 2\ CO_2 + 4\ H_2 \qquad (4.16)$$

$$C_6H_{12}O_6 \rightarrow CH_3CH_2COOH + 2\ CO_2 + 2\ H_2. \qquad (4.17)$$

Thus, the highest theoretical yields of H_2 are associated with acetate as the fermentation end product. In practice, however, high H_2 yields are usually associated with a mixture of acetate and butyrate fermentation products, and low H_2 yields are with associated propionate and reduced end-products (alcohols, lactic acid). *Clostridium pasteurianum*, *Clostridium butyricum*, and *Clostridium beijerinkii* are high H_2 producers, while *Clostridium propionicum* is a poor H_2 producer [75, 79, 80].

The amount of hydrogen production by dark fermentation highly depends on the pH value, hydraulic retention time (HRT) and gas partial pressure. For the optimal hydrogen production, pH should be maintained between 5 and 6 [23]. Hydrogen production from the bacterial fermentation of sugars has been examined in a variety of reactor systems. Hexose concentration has a greater effect on H_2 yields than the HRT. Flocculation also was an important factor in the performance of the reactor [40]. The hydrogen yield from sucrose by dark fermentation could be difficult to increase above 2.5-mol/mol hexose, owing to the formation of fatty acids during the fermentation process [81].

4.4.4 Two-Stage Process: Integration of Dark and Photofermentation

A combination of dark and photofermentation in a two-stage hybrid system could be expected to reach as close to the theoretical maximum production of 12 mol of H_2 (mol glucose)$^{-1}$ equivalent as possible, according to the following reactions [82]:

Stage I: Dark fermentation (facultative anaerobes)

$$C_6H_{12}O_6 + 2\ H_2O \rightarrow 2\ CH_3COOH + 2\ CO_2 + 4\ H_2 \qquad (4.18)$$

Stage II: Photofermentation (photosynthetic bacteria)

$$2\,CH_3COOH + 4\,H_2O \rightarrow 8\,H_2 + 4\,CO_2. \tag{4.19}$$

In the first step, biomass is fermented to acetate, carbon dioxide, and hydrogen by thermophilic dark fermentation, while in the second step, acetate is converted to hydrogen and carbon dioxide.

Starch, cellulose or hemicellulose content of wastes, carbohydrate rich food industry effluents or waste biological sludge can be further processed to convert the carbohydrates to organic acids and then to hydrogen gas by using proper bioprocessing technologies. Figure 4.4 shows schematic diagram for biohydrogen production from food industry wastewaters and agricultural wastes by two stages, anaerobic dark and photofermentations [83].

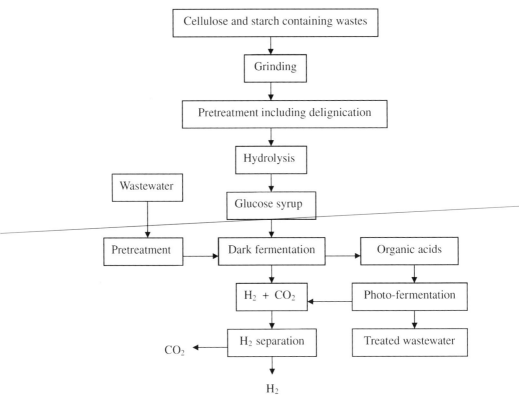

FIGURE 4.4 A schematic diagram for biohydrogen production from cellulose/starch containing agricultural wastes and food industry wastewaters. *Source*: Reproduced with permission from Kapdan and Kargi [83].

4.5 SUMMARY

Hydrogen produced from biorenewables is a promising sustainable energy carrier as an alternative to fossil fuels. Biomass-based hydrogen generation is particularly interesting for rural areas. The share of hydrogen from biomass in the automotive fuel market is expected to grow in the next decade. Hydrogen is currently more expensive than conventional energy sources. In the longer term, renewable sources will become increasingly more important for production of hydrogen, thereby lowering its production cost. Thermochemical (pyrolysis and gasification) and biological (biophotolysis, photofermentation, and dark fermentation) processes can be practically applied to produce hydrogen. Biomass gasification offers the earliest and most economical route for the production of renewable hydrogen. Steam reforming of natural gas and gasification of biomass could become the dominant technologies by the end of the twenty-first century.

REFERENCES

1. Demirbas, A. *Biohydrogen: For Future Engine Fuel Demands*, Trabzon, Springer, 2009.

2. Larsen H., Feidenhans'l R., Petersen LS. Hydrogen and its competitors, Risø energy Report 3, Risø National Laboratory, Roskilde, Denmark, November 2004.

3. Milne TA., Elam CC., Evans RJ. Hydrogen from biomass: State of the art and research challenges, Report for IEA, IEA/H2/TR-02/001. National Renewable Energy Laboratory, Golden, CO, 2002.

4. Patel AG., Maheshwari NK., Vijayan PK., Sinha RK. A study on sulfur-iodine (S-I) thermochemical water splitting process for hydrogen production from nuclear heat. In Proceedings of the Sixteenth Annual Conference of Indian Nuclear Society, Science behind Nuclear Technology, Mumbai, India, November 15–18, 2005.

5. Demirbas A. Yields of hydrogen of gaseous products via pyrolysis from selected biomass samples. *Fuel* **2001**, *80*, 1885–1891.

6. Wang D., Czernik S., Montane D., Mann M., Chornet E. Biomass to hydrogen via fast pyrolysis and catalytic steam reforming of the pyrolysis oil or its fractions. *Industrial & Engineering Chemistry Research* **1997**, *36*, 1507–1518.

7. Yan Q., Guo L., Lu Y. Thermodynamic analysis of hydrogen production from biomass gasification in supercritical water. *Energy Conversion and Management* **2006**, *47*, 1515–1528.

8. Swami SM., Chaudhari V., Kim DS., Sim SJ., Abraham MA. Production of hydrogen from glucose as a biomass simulant: Integrated biological and thermochemical approach. *Industrial & Engineering Chemistry Research* **2008**, *47*, 3645–3651.

9. Sricharoenchaikul V., Marukatat C., Atong T. Fuel production from physic nut (*Jatropha Curcas* L.) waste by fixed-bed pyrolysis process. In Proceedings of the 3rd Conference

on Energy Network of Thailand, ENETT2550-072, Bangkok, Thailand, May 23–25, 2007.

10. Das D., Veziroglu TN. Hydrogen production by biological processes: A survey of literature. *International Journal of Hydrogen Energy* **2001**, *26*, 13–28.

11. Nandi R., Sengupta S. Microbial production of hydrogen: An overview. *Critical Reviews in Microbiology* **1998**, *24*, 61–84.

12. Cheong DY., Hansen CL. Bacterial stress enrichment enhances anaerobic hydrogen production in cattle manure sludge. *Applied Microbiology and Biotechnology* **2006**, *72*, 635–643.

13. Ren N., Li J., Li B., Wang Y., Liu S. Biohydrogen production from molasses by anaerobic fermentation with a pilotscale bioreactor system. *International Journal of Hydrogen Energy* **2006**, *31*, 2147e57.

14. O-Thong S., Prasertsan P., Karakashev D., Angelidaki I. Thermophilic fermentative hydrogen production by the newly isolated *Thermoanaerobacterium thermosaccharolyticum* PSU-2. *International Journal of Hydrogen Energy* **2008**, *33*, 1204–1214.

15. Devrije T., Bakker RR., Budde MAW., Lai MH., Mars AE., Claassen PAM. Efficient hydrogen production from the lignocellulosic energy crop Miscanthus by the extreme thermophilic bacteria *Caldicellulosiruptor saccharolyticus* and *Thermotoga neapolitana*. *Biotechnology Biofuels* **2009**, *2*, 12.

16. Zheng H., O'Sullivan C., Mereddy R., Zeng RJ., Duke M., Clarke W. Production of biohydrogen using a membrane anaerobic reactor: Limitations due to diffusion. In Proceedings of the Environmental Research Event 2009, Noosa Heads, Queensland, Australia, May 10–13, 2009.

17. Kotay SM., Das D. Biohydrogen as a renewable energy resourced prospects and potentials. *International Journal of Hydrogen Energy* **2008**, *33*, 258–263.

18. Tao Y., Chen Y., Wu Y., He Y., Zhou Z. High hydrogen yield from a two-step process of dark- and photo-fermentation of sucrose. *International Journal of Hydrogen Energy* **2007**, *32*, 200–206.

19. Demirbas A. Thermochemical conversion of mosses and algae to gaseous products. *Energy Sources Part A. Recovery, Utilization, and Environmental Effects* **2009**, *31*, 746–753.

20. Balat H. Prospects of biofuels for a sustainable energy future: A critical assessment. *Energy Education Science and Technology, Part A* **2010**, *24*, 85–111.

21. Encinar JM., Beltran FJ., Ramiro A., Gonzalez JF. Pyrolysis/gasification of agricultural residues by carbon dioxide in the presence of different additives: Influence of variables. *Fuel Processing Technology* **1998**, *55*, 219–233.

22. Balat M. New biofuel production technologies. *Energy Education Science and Technology, Part A* **2009**, *22*, 147–161.

23. Ni M., Leung DYC., Leung MKH., Sumathy K. An overview of hydrogen production from biomass. *Fuel Processing Technology* **2006**, *87*, 461–472.

24. Demirbas A. Partial hydrogenation effect of moisture contents on the combustion oils from biomass pyrolysis. *Energy Sources. Part A. Recovery, Utilization, and Environmental Effects* **2008**, *30*, 508–515.

25. Demirbas A. Hydrogen-rich gases from biomass via pyrolysis and air–steam gasification. *Energy Sources. Part A. Recovery, Utilization, and Environmental Effects* **2009**, *31*, 1728–1736.

26. Demirbas A. Gaseous products from biomass by pyrolysis and gasification: Effects of catalyst on hydrogen yield. *Conversion and Management* **2002**, *43*, 897–909.

27. Demirbas A. Hydrogen production via pyrolytic degradation of agricultural residues. *Energy Sources. Part A. Recovery, Utilization, and Environmental Effects* **2005**, *27*, 769–775.

28. Berman A., Karn RK., Epstein M. Kinetics of steam reforming of methane on Ru/Al_2O_3 catalyst promoted with Mn oxides. *Applied Catalysis A: General* **2005**, *282*, 73–83.

29. Galdamez JM., Garcia L., Bilbao R. Hydrogen production by steam reforming of bio-oil using coprecipitated Ni–Al catalysts. Acetic acid as a model compound. *Energy Fuels* **2005**, *19*, 1133–1142.

30. Vagia EC., Lemonidou AA. Thermodynamic analysis of hydrogen production via steam reforming of selected components of aqueous bio-oil fraction. *International Journal of Hydrogen Energy* **2007**, *32*, 212–223.

31. Wu C., Yan Y., Li T., Qi W. Preparation of hydrogen through catalytic steam reforming of bio-oil. *The Chinese Journal of Process Engineering* **2007**, *7*, 1114–1119.

32. Evans RJ., Chornet E., Czernik S., Feik C., French R., Phillips S., et al. Renewable hydrogen production by catalytic steam reforming of peanut shells pyrolysis products. *American Chemical Society: Division of Fuel Chemistry* **2002**, *47*, 757–758.

33. Nath K., Das D. Hydrogen from biomass. *Current Science* **2003**, *85*, 265–271.

34. Turner J., Sverdrup G., Mann MK., Maness PC., Kroposki B., Ghirardi M., et al. Renewable hydrogen production. *International Journal of Energy Research* **2008**, *32*, 379–407.

35. Demirbas MF. Technological options for producing hydrogen from renewable resources. *Energy Sources. Part A. Recovery, Utilization, and Environmental Effects* **2006**, *28*, 1215–1223.

36. Demirbas A. Hydrogen production from carbonaceous solid wastes by steam reforming. *Energy Sources. Part A. Recovery, Utilization, and Environmental Effects* **2008**, *30*, 924–931.

37. Yoon SJ., Choi YC., Lee JG. Hydrogen production from biomass tar by catalytic steam reforming. *Energy Conversion and Management* **2010**, *51*, 42–47.

38. Stassen HEM., Knoef HAM. Small scale gasification systems, Biomass Technology Group, University of Twente, The Netherlands, 1993.

39. Balat M. Sustainable transportation fuels from biomass materials. *Energy Education Science and Technology* **2006**, *17*, 83–103.

40. Demirbas A. Converting biomass derived synthetic gas to fuels via Fisher–Tropsch synthesis. *Energy Sources. Part A. Recovery, Utilization, and Environmental Effects* **2007**, *29*, 1507–1512.

41. Wei L., Xu S., Zhang L., Liu C., Zhu H., Liu S. Steam gasification of biomass for hydrogen-rich gas in a free-fall reactor. *International Journal of Hydrogen Energy* **2007**, *32*, 24–31.

42. Demirbas A. Thermochemical conversion of hazelnut shell to gaseous products for production of hydrogen. *Energy Sources. Part A. Recovery, Utilization, and Environmental Effects* **2005**, *27*, 339–347.

43. Kriengsak SN., Buczynski R., Gmurczyk J., Gupta AK. Hydrogen production by high-temperature steam gasification of biomass and coal. *Environmental Engineering Science* **2009**, *26*, 739–744.

44. Demirbas MF. Producing hydrogen from biomass via non-conventional processes. *Energy, Exploration & Exploitation* **2004**, *22*, 225–233.

45. Demirbas MF. Hydrogen from various biomass species via pyrolysis and steam gasification processes. *Energy Source A* **2006**, *28*, 245–252.

46. Wu W., Kawamoto K., Kuramochi H. Hydrogen-rich synthesis gas production from waste wood via gasification and reforming technology for fuel cell application. *Journal of Material Cycles and Waste Management* **2006**, *8*, 70–77.

47. Demirbas A. Hydrogen production from biomass by gasification process. *Energy Sources. Part A. Recovery, Utilization, and Environmental Effects* **2002**, *24*, 59–68.

48. Dupont C., Boissonnet G., Seiler JM., Gauthier P., Schweich D. Study about the kinetic processes of biomass steam gasification. *Fuel* **2007**, *86*, 32–40.

49. Maschio G., Lucchesi A., Stoppato G. Production of syngas from biomass. *Bioresource Technology* **1994**, *48*, 119–126.

50. Demirbas A. Pyrolysis and steam gasification processes of black liquor. *Energy Conversion and Management* **2002**, *43*, 877–884.

51. Lv P., Chang J., Wang T., Fu Y., Chen Y. Hydrogen-rich gas production from biomass catalytic gasification. *Energy Fuels* **2004**, *18*, 228–233.

52. Demirbas A., Caglar A. Catalytic steam reforming of biomass and heavy oil residues to hydrogen. *Energy Education Science and Technology* **1998**, *1*, 45–52.

53. Luo S., Xiao B., Guo X., Hu Z., Liu S., He M. Hydrogen-rich gas from catalytic steam gasification of biomass in a fixed bed reactor: Influence of particle size on gasification performance. *International Journal of Hydrogen Energy* **2009**, *34*, 1260–1264.

54. Demirbas A. Characterization of products from two lignite samples by supercritical fluid extraction. *Energy Sources. Part A. Recovery, Utilization, and Environmental Effects* **2004**, *26*, 933–939.

55. Demirbas A. Diesel-like fuel from tallow by pyrolysis and supercritical water liquefaction. *Energy Sources. Part A. Recovery, Utilization, and Environmental Effects* **2009**, *31*, 824–830.

56. Lu YJ., Guo LJ., Ji CM., Zhang XM., Hao XH., Yan QH. Hydrogen production by biomass gasification in supercritical water: A parametric study. *International Journal of Hydrogen Energy* **2006**, *31*, 822–831.

57. Guo LJ., Lu YJ., Zhang XM., Ji CM., Guan Y., Pei AX. Hydrogen production by biomass gasification in supercritical water: A systematic experimental and analytical study. *Catalysis Today* **2007**, *129*, 275–286.

58. Yan Q., Guo L., Lu Y. Thermodynamic analysis of hydrogen production from biomass gasification in supercritical water. *Energy Conversion and Management* **2006**, *47*, 1515–1528.

59. Demirbas A. *Biohydrogen: For Future Engine Fuel Demands*, Springer-Verlag, London, 2009.

60. Demirbas A. Biorefineries: Current activities and future developments. *Energy Conversion and Management* **2009**, *50*, 2782–2801.

61. Wang D., Czernik S., Montane D., Mann M., Chornet E. Biomass to hydrogen via fast pyrolysis and catalytic steam reforming of the pyrolysis oil or its fractions. *Industrial & Engineering Chemistry Research* **1997**, *36*, 1507–1518.

62. Das D., Veziroglu TN. Hydrogen production by biological processes: A survey of literature. *International Journal of Hydrogen Energy* **2001**, *26*, 13–28.

63. Demirbas A. Biohydrogen generation from organic waste. *Energy Sources. Part A. Recovery, Utilization, and Environmental Effects* **2008**, *30*, 475–482.

64. Li J., Li B., Zhu G., Ren N., Bo L., He J. Hydrogen production from diluted molasses by anaerobic hydrogen producing bacteria in an anaerobic baffled reactor (ABR). *International Journal of Hydrogen Energy* **2007**, *32*, 3274–3283.

65. Zhi X., Yang H., Yuan Z., Shen J. Bio-hydrogen production of anaerobic bacteria in reverse micellar media. *International Journal of Hydrogen Energy* **2008**, *33*, 4747–4754.

66. Nath K., Das D. Biohydrogen production as a potential energy resource: Present state-of-art. *Journal of Scientific and Industrial Research* **2004**, *63*, 729–738.

67. Das D., Khanna N., Veziroglu TN. Recent developments in biological hydrogen production processes. *Chemical Industry and Chemical Engineering Quarterly* **2008**, *14*, 57–67.

68. Das D., Dutta T., Nath K., Kotay SM., Das AK., Veziroglu TN. Role of Fe-hydrogenase in biological hydrogen production. *Current Science* **2006**, *90*, 1627–1637.

69. Benemann JR. Hydrogen production by microalgae. *Journal of Applied Phycology* **2000**, *12*, 291–300.

70. Melis A., Zhang L., Forestier M., Ghirardi ML., Seibert M. Sustained photobiological hydrogen gas production upon reversible inactivation of oxygen evolution in the green alga *Chlamydomonas reinhardtii*. *Plant Physiology* **2000**, *122*, 127–136.

71. Resnick RJ. The economics of biological methods of hydrogen production. Master thesis, Massachusetts Institute of Technology, Sloan School of Management, Management of Technology Program; 2004.

72. Manish S., Banerjee R. Comparison of biohydrogen production processes. *International Journal of Hydrogen Energy* **2008**, *33*, 279–286.

73. Hallenbeck PC., Benemann JR. Biological hydrogen production; fundamentals and limiting processes. *International Journal of Hydrogen Energy* **2002**, *27*, 1185–1193.

74. Balat M. Production of hydrogen via biological processes. *Energy Sources. Part A. Recovery, Utilization, and Environmental Effects* **2009**, *31*, 1802–1812.

75. Levin DB., Pitt L., Love M. Biohydrogen production: Prospects and limitations to practical application. *International Journal of Hydrogen Energy* **2004**, *29*, 173–185.

76. Fedorov AS., Tsygankov AA., Rao KK., Hall DO. Hydrogen photoproduction by *Rhodobacter sphaeroides* immobilised on polyurethane foam. *Biotechnology Letters* **1998**, *20*, 1007–1009.

77. Gadhamshetty V., Sukumaran A., Nirmalakhandan N., Myint MT. Photofermentation of malate for biohydrogen production: A modeling approach. *International Journal of Hydrogen Energy* **2008**, *33*, 2138–2146.

78. Ren N., Li J., Li B., Wang Y., Liu S. Biohydrogen production from molasses by anaerobic fermentation with a pilot-scale bioreactor system. *International Journal of Hydrogen Energy* **2006**, *31*, 2147–2157.

79. Hawkes FR., Dinsdale R., Hawkes DL., Hussy I. Sustainable fermentative hydrogen production: Challenges for process optimisation. *International Journal of Hydrogen Energy* **2002**, *27*, 1339–1347.

80. Nath K., Das D. Improvement of fermentative hydrogen production: Various approaches. *Applied Microbiology and Biotechnology* **2004**, *65*, 520–529.

81. Tao Y., Chen Y., Wu Y., He Y., Zhou Z. High hydrogen yield from a two-step process of dark- and photo-fermentation of sucrose. *International Journal of Hydrogen Energy* **2007**, *32*, 200–206.

82. Kotay SM., Das D. Biohydrogen as a renewable energy resource: Prospects and potentials. *International Journal of Hydrogen Energy* **2008**, *33*, 258–263.

83. Kapdan IK., Kargi F. Bio-hydrogen production from waste materials. *Enzyme and Microbial Technology* **2006**, *38*, 569–582.

5

Established Methods Based on Compression and Cryogenics

5.1 BASIC ISSUES ABOUT HYDROGEN STORAGE

Hydrogen storage is one of the major challenges in the development of hydrogen as a fuel for widespread applications. In this chapter, we will cover traditional or well-established techniques for hydrogen storage based on high pressure compression or low temperature liquefaction. Other newer methods under research and development will be covered in Chapters 6 and 7. We will first introduce energy content for fuels.

The energy stored in hydrogen or other fuels can be expressed either on a weight basis (mass energy density or gravimetric capacity) or on a volume basis (volumetric energy density or volumetric capacity). When hydrogen reacts with oxygen, water (vapor or liquid) is formed and energy is released,

$$H_2 + \frac{1}{2}O_2 = H_2O \quad \Delta H = 241.826 \, kJ \cdot mol^{-1} (lower \ heating \ value). \quad (5.1)$$

Thus, if 1 mol of hydrogen is burned with 100% energy conversion efficiency, 241.826 kJ of energy should be released. Since the molar mass of hydrogen is $M = 2.02 \times 10^{-3}$ kg·mol^{-1}, the mass energy density of pure hydrogen is

$$\rho_M^0 = \frac{\Delta H}{M} = 119.716 \, MJ \cdot kg^{-1}. \quad (5.2)$$

Hydrogen Generation, Storage, and Utilization, First Edition. Jin Zhong Zhang, Jinghong Li, Yat Li, and Yiping Zhao.
© 2014 John Wiley & Sons, Inc. Published 2014 by John Wiley & Sons, Inc.

At 1 atm and 298.15 K (25°C), the volume occupied by 1 mole of H_2 is $V_M = 24.46$ L, and the volumetric energy density of H_2 is

$$\rho_V = \frac{\Delta H}{V_M} = 9.89 \text{ MJ} \cdot \text{m}^{-3}. \tag{5.3}$$

Similar definition can be used for other fuels, such as methane, propane, and gasoline. In fact, hydrogen has the highest mass energy density among all the chemical fuels, but almost the lowest volumetric energy density beside wood. Taken gasoline for example, its mass energy density is 45.7 $MJ \cdot kg^{-1}$ and volumetric energy density is 34,600 $MJ \cdot m^{-3}$. Although gasoline has smaller mass energy density, it has the highest volumetric energy density that makes it really useful. For practical application, for example, it takes 10 gal of gasoline for a light-duty vehicle to drive around 300 mi. If hydrogen is used, one needs to burn a tank of 3495 gal of hydrogen to drive similar distance. It is practically impossible to use such a big volume of H_2 to drive a commercial vehicle. Therefore, one critical issue in using hydrogen is to find new methods to improve the volumetric energy density of hydrogen while keeping the mass energy density high, that is, to compress the large volumes of the hydrogen gas.

One of the easiest ways to increase the volumetric energy density is to compress the hydrogen or to liquefy hydrogen at low temperature. Liquid hydrogen has a mass density of 70.8 $kg \cdot m^{-3}$ (at -253°C). This gives a volumetric energy density of 8.495×10^3 $MJ \cdot m^{-3}$, which is about 860 times higher than that of hydrogen gas at ambient conditions. This storage method is based on changing the physical state of hydrogen, and usually requires extra accessories, such as robust containers, valves, regulators, piping, mounting brackets, insulation, added cooling capacity, and thermal management components.

An alternative way to store hydrogen is to use hydrogen storage materials. For some solid materials, the interaction of hydrogen atoms with the atoms in the materials may be much stronger than the hydrogen–hydrogen interaction. Therefore, hydrogen atoms could bind more closely together inside the solid structures, generating much higher hydrogen density under ambient conditions, or hydrogen could chemically react with the solid material to form hydrides with relatively weak hydrogen bonds or metastable states. Those materials could significantly improve the volumetric energy density of hydrogen.

For hydrogen storage system, the mass of the energy storage system not only includes the mass of hydrogen m_H, but also should account for mass of containers or storage materials, m_s. Thus, the effective mass energy density ρ_M of a hydrogen storage system can be expressed as,

$$\rho_M = \frac{\Delta H}{m_H + m_s} \times \frac{m_H}{M} = \frac{\Delta H}{M} \times \frac{m_H}{m_H + m_s} = \frac{m_H}{m_H + m_s} \rho_M^0. \qquad (5.4)$$

Since $\rho_M^0 = 119.716 \, \text{MJ} \cdot \text{kg}^{-1}$ is a constant, the effective mass energy density is proportional to the hydrogen's mass percentage

$$\frac{m_H}{m_H + m_s}(\%),$$

in the storage system or materials, which means for a hydrogen storage system or materials, the mass energy density is only a fraction of that in pure hydrogen. Therefore, one can use the hydrogen mass percentage to describe the gravimetric capacity of hydrogen storage systems. Similarly, the effective volumetric energy density ρ_V of a hydrogen storage system or material can be expressed as,

$$\rho_V = \frac{\Delta H}{V} \times \frac{m_H}{M} = \rho_M^0 \frac{m_H}{V}, \qquad (5.5)$$

where V is the total volume of the storage system or materials. Thus, one can use the effective hydrogen density

$$\frac{m_H}{V}(\text{g} \cdot \text{L}^{-1}),$$

to describe the volumetric capacity of hydrogen storage systems.

For onboard hydrogen applications, a vehicle needs to carry 5–13 kg hydrogen in order to make a drive distance great than 300 mi. With the confinement of vehicle design, this imposes many requirements for hydrogen storage systems. U.S. Department of Energy (DOE) has set several targets for onboard hydrogen storage systems for light-duty vehicles [1]. Its ultimate target is 7.5% gravimetric capacity and 70 g·L^{-1} volumetric capacity. The target for year 2017 is 5.5% and 40 g·L^{-1} gravimetric and volumetric capacities. Figure 5.1 shows a summary of the different hydrogen storage systems in use or under development in terms of their gravimetric and volumetric capacities as compared with DOE's targets. Clearly, the physical means to store hydrogen, for example, the compressed hydrogen and liquid hydrogen systems, are very close to the DOE targets.

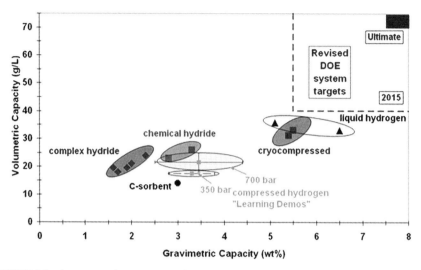

FIGURE 5.1   A summary of current status of hydrogen storage technologies in terms of weight, volume, and cost. These values are estimates from storage system developers and the R&D community and will be continuously updated by DOE as new technological advancements take place. *Source*: Reproduced with permission from http://www1.eere.energy.gov/hydrogenandfuelcells/storage/tech_status.html [2]. (See color insert.)

5.2   HIGH PRESSURE COMPRESSION

A common and simple method for storing and transporting hydrogen is to compress hydrogen into a fixed volume (in a metal cylinder or tank) at high pressure, so that the mass density of hydrogen will be increased. If under high pressure P, the mass of compressed hydrogen is m_H, the tank volume and mass are V_{tank} and m_{tank}, the effective gravimetric and volumetric capacities are

$$\rho'_M = \frac{m_H}{m_H + m_{tank}} \text{ and } \rho'_V = \frac{m_H}{V_{tank}}.$$

The parameters P, V_{tank}, and m_H are linked by the equation of state of hydrogen, $f(P, V_{tank}, T) = 0$. The simplest approximation is the law of an ideal gas,

$$PV_{tank} = nRT. \tag{5.6}$$

where

$$n = \frac{m_H}{M},$$

and R is the gas constant. If Equation (5.6) is valid, then

$$\rho_V' = \frac{PM}{RT},\qquad(5.7)$$

that is, the volumetric capacity increases linearly with compression pressure P. However, the compressed hydrogen gas cannot be treated as an idea gas due to strong intermolecular interactions, and the equation of state can be approximated as

$$PV_{\text{tank}} = nZRT,\qquad(5.8)$$

where Z is called the compressibility factor, and is a function of both temperature T and pressure P, as shown in Figure 5.2. The volumetric capacity can be estimated as

$$\rho_V' = \frac{PM}{ZRT}.\qquad(5.9)$$

Depending on the compression temperature, ρ_V' may not be a monotonic function of P. Table 5.1 shows the calculated ρ_V' values at 20°C for a 150-L

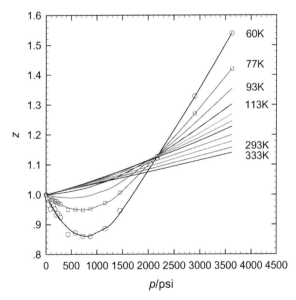

FIGURE 5.2   Compressibility factor of hydrogen. *Source*: Reproduced with permission from Zhou and Zhou [3].

TABLE 5.1   Hydrogen Compressibility Factor (Z) at 20°C and Corresponding Volumetric Capacity ρ'_v [4]

P(MPa)	0.1013	5	10	20	30	35	40	50	70	100
Z	1	1.032	1.065	1.132	1.201	1.236	1.272	1.344	1.489	1.702
m_H (kg) in 150-L tank)	0.0126	0.603	1.17	2.20	3.11	3.52	3.91	4.63	5.85	7.31
ρ'_v (g·L^{-1})	0.084	4.02	7.79	14.65	20.71	23.48	26.08	30.85	38.98	48.72

tank at different compression pressure. The corresponding Z-factor is also given in Table 5.1 [3]. For a standard 150-L compression tank, only when the pressure increases to more than 50 MPa can the amount of hydrogen stored in the compressed gas tank meet the requirement by DOE.

The gravimetric capacity can be expressed as

$$\rho'_M = \frac{1}{1 + m_{\text{tank}} / m_H} = \frac{1}{1 + m_{\text{tank}} ZRT / MPV_{\text{tank}}}. \tag{5.10}$$

The mass of hydrogen m_H stored in standard tank is calculated and listed in Table 5.1. Clearly, the gravimetric capacity is determined by the mass of the compressed tank. To meet the DOE's 2017 target, the mass of the tank $m_{\text{tank}} \leq 100.5$ kg if the hydrogen can be compressed at 70 MPa.

Compressed hydrogen is stored in thick-walled tanks made of high strength materials to ensure durability and safety. The standard compressed tanks use pressure of 10–20 MPa, and are usually made of heavy steel- or aluminum-lined steel. They cannot hold enough hydrogen for onboard applications, and have a significantly low gravimetric and volumetric capacity (see Table 5.1). The current trend to replace the standard gas cylinders is to use lightweight composite fiber tanks. The state-of-art advanced lightweight storage system is based on the designed philosophy of TriShield™ cylinder (QUANTUM Technologies WorldWide, Inc.) as shown in Figure 5.3. The system is comprised of a seamless, one-piece, permeation-resistant, cross-linked ultra-high molecular weight polymer liner that is overwrapped with multiple layers of carbon fiber/epoxy laminate and a proprietary external protective layer for impact resistance. Currently, QUANTUM has a commercial 129-L lightweight H_2 cylinder working under 70 MPa. The weight of the cylinder is 92 kg, and stored useful hydrogen is 5 kg [6].

Besides the hydrogen content, another critical issue associated with high pressure hydrogen storage is the compression. To compress hydrogen requires

Impact-Resistant
Dome
· Light-weight
· Energy Absorbing
· Cost-Competitive

Manual Valve, or
Electrical Valve or
In-Tank Regulator

Ploymer Liner
· Light-weight
· Corrosion resistant (hydrogen
  embrittlement)
· Permeation barrier
· Cost-Competitive
· Flexible in Size

Carbon-fiber
Reinforced Shell
· Corrosion resistant
  (acids, bases)
· Fatigue/Creep/Relaxation
  resistant
· Light-weight

Reinforced External
Protective Shell
· Gunfire safety
· Impact safety
· Cut/Abrasion
  Resisitance

FIGURE 5.3 TriShield™ Tank construction. *Source*: Reproduced with permission from http://www1 .eere.energy.gov/hydrogenandfuelcells/pdfs/32405b27.pdf [5].

external electrical energy. The actual work done to compress a gas is usually between the theoretical estimations of an isothermal and an adiabatic compression process. The work W required to compress the gas is

$$W = \int_{V_1}^{V_2} PdV. \tag{5.11}$$

For an isothermal compression, the temperature of hydrogen is assumed not to change during the process, and for an ideal gas, the work W_T is

$$W_T = nRTln\frac{P_2}{P_1}. \tag{5.12}$$

For a nonideal gas with a Z-factor (Eq. 5.8),

$$W_T \approx nZRTln\frac{P_2}{P_1}, \tag{5.13}$$

where V_1 and P_1 are the initial uncompressed volume and pressure, while V_2 and P_2 are the final compressed volume and pressure of hydrogen. For an ideal gas in a adiabatic compression (no heat exchange with the environment), it is well known that [7]

$$PV^\gamma = P_1V_1^\gamma = \text{constant}, \tag{5.14}$$

and

$$\left(\frac{V_1}{V_2}\right)^{\gamma-1} = \left(\frac{P_2}{P_1}\right)^{\frac{\gamma-1}{\gamma}}, \tag{5.15}$$

where $\gamma = C_p/C_v$ is the specific heat ratio, and C_p and C_v are the specific heats of hydrogen. For an ideal hydrogen gas, $\gamma = 1.4$. The work W_Q done in the adiabatic process is

$$W_Q = \frac{\gamma}{\gamma-1} nRT_1 \left[\left(\frac{P_2}{P_1}\right)^{\frac{\gamma-1}{\gamma}} - 1\right]. \tag{5.16}$$

Usually, the compression process is carried out under several stages, and the total work required for an l-stage process is

$$W_Q = \frac{\gamma}{\gamma-1} nRT_1 \sum_{i=1}^{l} \left[\left(\frac{P_{i+1}}{P_i}\right)^{\frac{\gamma-1}{\gamma}} - 1\right]. \tag{5.17}$$

Multiple stage compression could significantly reduce the required energy. For example, for a two-stage compression, if the initial and final pressures are fixed at P_1 and P_2, when the intermediate pressure $P_i = \sqrt{P_1 P_2}$, Equation (5.17) gives the minimum work required,

$$W_Q = \frac{2\gamma}{\gamma-1} nRT_1 \left[\left(\frac{P_2}{P_1}\right)^{\frac{\gamma-1}{2\gamma}} - 1\right]. \tag{5.18}$$

If 1 mole of hydrogen is compressed from 1 atm at 20°C, Figure 5.4 shows the comparison of the works W_Q required for a one-stage and a two-stage compression as a function of the final pressure P_2. To compress the hydrogen to 70 MPa, the two-stage process only requires $\sim$56% of energy of the one-stage process. The higher the final pressure, the more the energy saved. Similarly, a three-stage process, with intermediate pressure

$$P_{i1} = P_1^{2/3} P_2^{1/3}, P_{i2} = P_1^{1/3} P_2^{2/3} \quad \text{and} \quad W_Q = \frac{3\gamma}{\gamma-1} nRT_1 \left[\left(\frac{P_2}{P_1}\right)^{\frac{\gamma-1}{3\gamma}} - 1\right],$$

can further reduce the required work (Fig. 5.4). Thus, with the increase of number of compression stages, the required energy is reduced. But no significant energy efficiency gain is expected when $l > 3$.

When the pressure becomes higher, the behavior of hydrogen gas deviates from that of an ideal gas, and the compression process can be better

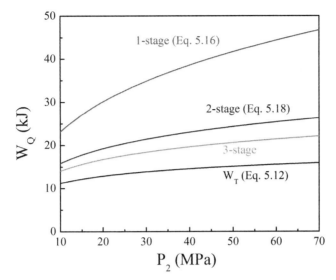

FIGURE 5.4   Energy required to compress 1 mole hydrogen from 1 atm at 20°C for a one-stage, a two-stage, and a three-stage compression process. The lowest curve show W_T calculated using Equation (5.12) for the ideal gas case. (See color insert.)

approximated by a polytropic process. The index γ in Equation. (5.16) and Equation (5.17) can be replaced by the polytropic index β ($\geq\gamma$) with

$$\beta = \frac{\eta\gamma}{1+\eta\gamma-\gamma}, \tag{5.19}$$

while keeping the similar formula for Equation (5.14), Equation (5.15), Equation (15.16), Equation (5.17), and Equation (5.18). Here, η is the polytropic efficiency.

There are other issues associated with the compressed hydrogen storage, in particular, cost and safety. The cost includes that for the lightweight tank and for the compressor as well as the energy required for compression. Novel compression technique with low cost and low energy requirement is also a challenge issue and active topic for high pressure hydrogen storage. For example, recently, the electrochemical hydrogen pumps based on the development of polymer electrolyte membrane fuel cell has been proposed and studied extensively [8]. An example of polymer electrolyte hydrogen pump (PEHP) is shown in Figure 5.5 [9]. It consists of a proton-conducting polymer electrolyte sandwiched between two porous electrodes. Gas containing H_2 is fed to the anode. Under an applied voltage, H_2 is oxidized into protons and electrons. The protons can transport across the polymer electrolyte

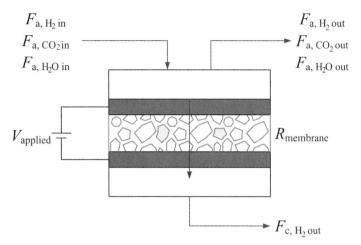

FIGURE 5.5 An example of the polymer electrolyte hydrogen pump (PEHP). *Source*: Reproduced with permission from Abdulla et al. [9]. (See color insert.)

membrane to the cathode, where they are reduced back to H_2. The pumping is done at low potentials, and only hydrogen is pumped across the electrolyte and membrane. The PEHP acts both as a separator unit and as a pump.

5.3 LIQUID HYDROGEN

Liquid hydrogen has a much higher volumetric capacity compared with gas hydrogen as shown in Section 5.1. It is expected that if hydrogen is stored in liquid phase, it will maintain high energy content, which could meet the DOE target. In fact, liquid hydrogen has been used in some special applications such as the Space Shuttle. However, liquid hydrogen only exists in a very narrow temperature and pressure range as shown in the phase diagram in Figure 5.6: between the triple point 13.8 K (at this temperature, gas, liquid, and solid phases coexist) and critical point 32.97 K (the heat of vaporization is zero at and beyond this temperature). Hydrogen has the second lowest boiling point (transition temperature from liquid to gas), 20.28 K, and melting point (transition temperature from solid to liquid), 14.01 K, of all substances, second only to helium. The boiling point is a critical parameter since it defines the temperature to be cooled in order to store and use the fuel as a liquid. Thus, liquid hydrogen requires cryogenic storage and boils around 20.28 K ($-252.87°C$ or $-423.17°F$). Cooling to such low temperature requires a fairly amount of energy, and storing the liquid hydrogen needs specific containers and management.

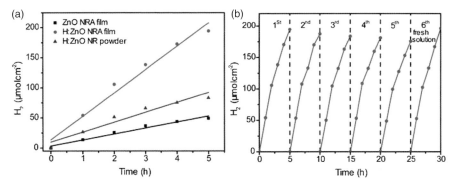

FIGURE 3.1   (a) Photocatalytic hydrogen generation rate collected for ZnO nanorod arrays (NRA) film, hydrogen-treated ZnO (H:ZnO) NRA film, and H:ZnO nanorod (NR) powder in a solution containing 0.1 M Na2SO3 and 0.1 M Na2S under white light irradiation. (b) Cycling performance of H:ZnO NRA films. Source: Reproduced with permission from Lu et al. [24].

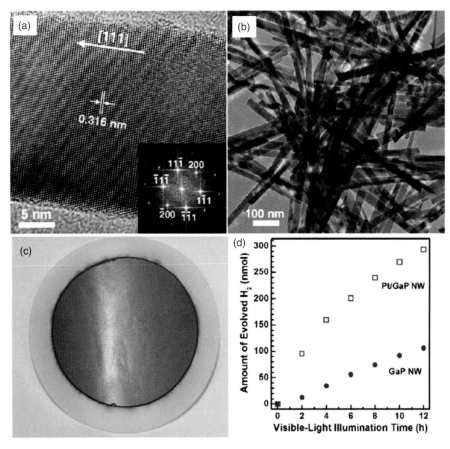

FIGURE 3.2   (a–b) TEM images of GaP nanowires. The inset in (a) shows the indexed FFT pattern of the image, indicating the wire is a single crystal with a growth axis of [111] direction. (c) Photograph of a large GaP nanowire membrane on a PVDF filter membrane. *Source*: Reproduced with permission from Sun et al. [18].

Hydrogen Generation, Storage, and Utilization, First Edition. Jin Zhong Zhang, Jinghong Li, Yat Li, and Yiping Zhao.
© 2014 John Wiley & Sons, Inc. Published 2014 by John Wiley & Sons, Inc.

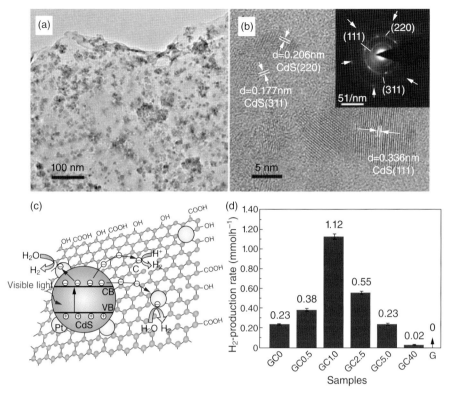

FIGURE 3.3 (a,b) TEM images of graphene sheet decorated with CdS clusters. Inset: SAED pattern collected at the composite structure. (c) Schematic illustration of the charge separation and transfer in the graphene-CdS system under visible light. (d) Comparison of the visible light photocatalytic activity of graphene–CdS systems with different graphene loading for the H_2 production using 10 vol% lactic acid aqueous solution as a sacrificial reagent and 0.5 wt% Pt as a co-catalyst. *Source*: Reproduced with permission from Li et al. [30].

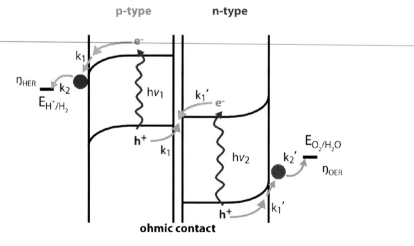

FIGURE 3.4 Schematic energy diagram of PEC water splitting with (a) photoanode, (b) photocathode, and (c) n-type photoanode and p-type photocathode. Source: Reproduced with permission from Liu et al. [31]. a PVDF filter membrane. *Source*: Reproduced with permission from Sun et al. [18].

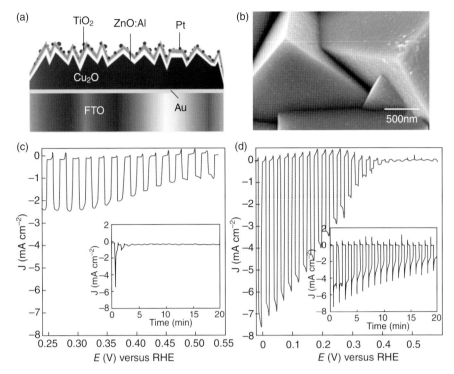

FIGURE 3.5 (a) Schematic presentation of the electrode structure. (See text for full caption.)

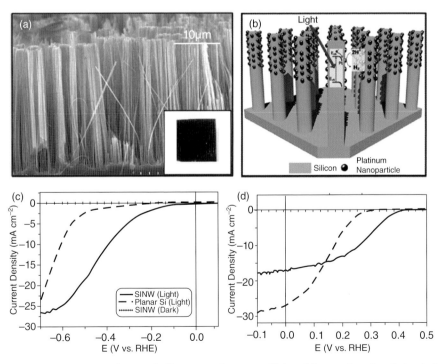

FIGURE 3.6 (A) SEM image of silicon nanowire arrays fabricated by metal-catalyzed chemical etching; inset is the photograph of ∼10 mm × 10 mm silicon nanowire array sample with low reflection. (See text for full caption.)

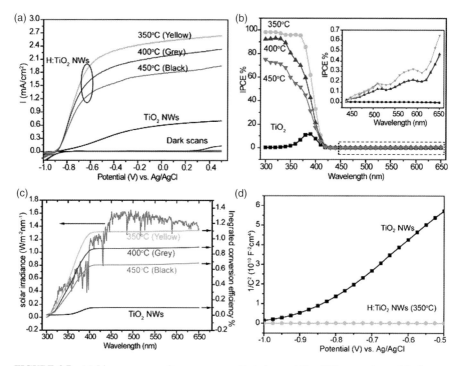

FIGURE 3.7 (a) Linear sweep voltammograms collected on pristine TiO_2 nanowire and hydrogen-treated TiO_2 ($H:TiO_2$) nanowires annealed at temperature of 350, 400, and 450°C. (b) IPCE spectra of pristine TiO_2 and $H:TiO_2$ nanowires. The inset is the magnified IPCE spectra that highlighted in the dashed box. (c) Simulated solar-to-hydrogen efficiencies for the pristine TiO_2 and $H:TiO_2$ samples as a function of wavelength, by integrating their IPCE spectra collected at −0.6 V versus Ag/AgCl with a standard AM 1.5G solar spectrum. (d) Mott–Schottky plots collected at a frequency of 5 kHz in the dark for pristine TiO_2 and $H:TiO_2$ nanowire. *Source*: Reproduced with permission from Wang et al. [11].

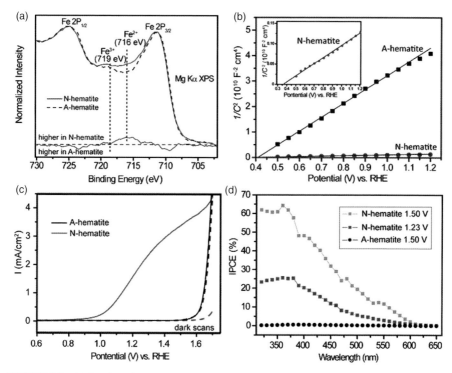

FIGURE 3.8 (a) Overlay of Fe 2p XPS spectra of air annealed hematite (denoted as: A-hematite) and oxygen-deficient hematite (denoted as: N-hematite), together with their different spectrums. The dashed lines highlight the satellite peaks of Fe^{2+} and Fe^{3+}. (b) Mott–Schottky plots measured for A-hematite and N-hematite. Inset: magnified Mott-schottky plot of N-hematite. (c) Linear sweep voltammograms collected on A-hematite and N-hematite under a simulated solar light of 100 $mW \cdot cm^{-2}$ and dark condition with a scan rate of 10 $mV \cdot s^{-1}$. (d) The corresponding IPCE spectra for A-hematite and N-hematite collected at potentials of 1.23 and 1.5V versus RHE. *Source*: Reproduced with permission from Ling et al. [46].

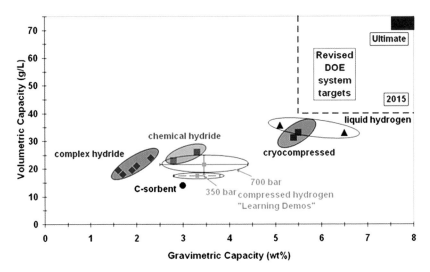

FIGURE 5.1 A summary of current status of hydrogen storage technologies in terms of weight, volume, and cost. These values are estimates from storage system developers and the R&D community and will be continuously updated by DOE as new technological advancements take place. *Source*: Reproduced with permission from http://www1.eere.energy.gov/hydrogenandfuelcells/storage/tech_status.html [2].

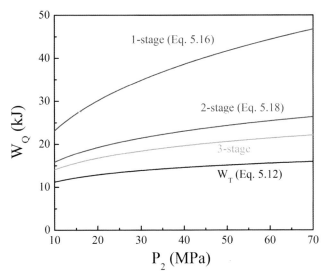

FIGURE 5.4 Energy required to compress 1 mole hydrogen from 1 atm at 20°C for a one-stage, a two-stage, and a three-stage compression process. The lowest curve show W_T calculated using Equation (5.12) for the ideal gas case.

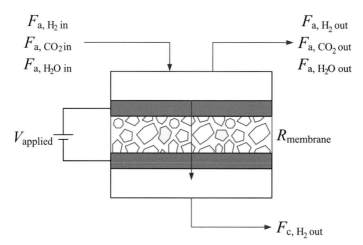

FIGURE 5.5   An example of the polymer electrolyte hydrogen pump (PEHP). *Source*: Reproduced with permission from Abdulla et al. [9].

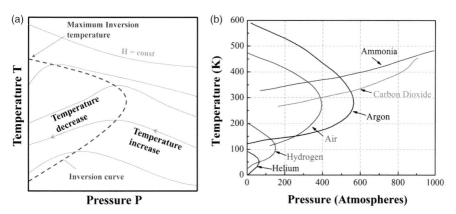

FIGURE 5.7   The illustration of the JT inversion curve (reproduced from Barron [11]) and the JT inversion curves for some conventional gases. *Source*: Reproduced with permission from Flynn [12].

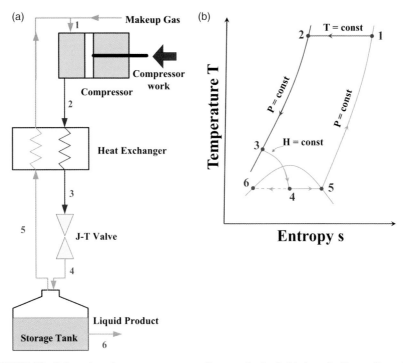

FIGURE 5.8 Schematic and temperature-entropy diagram of a simple Linde cycle. *Source*: Reproduced with permission from Barron [11].

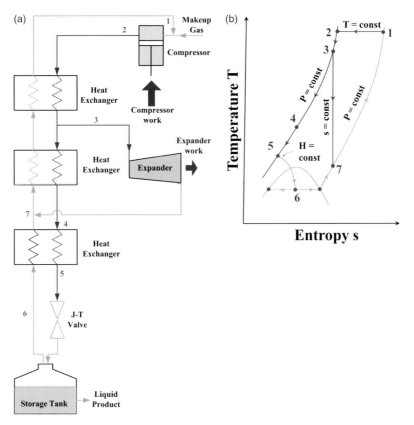

FIGURE 5.9 Schematic and temperature-entropy diagram of a simple Claude cycle. Reproduced with permission from Barron [11].

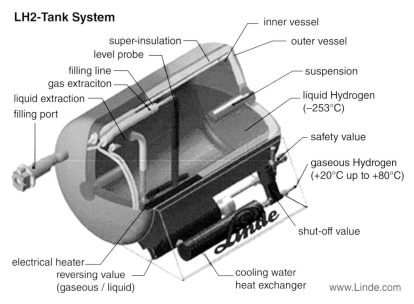

FIGURE 5.10 Schematic illustration of a representative cryogenic vessel. Source: "Hydrogen storage: state-of-the-art and future perspective," http://publications.jrc.ec.europa.eu/repository/bitstream/111111111/6013/1/EUR%2020995%20EN.pdf.

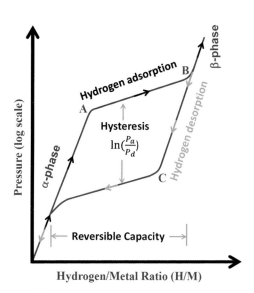

FIGURE 6.2 The typical PCT curves for the hydrogenation and dehydrogenation of a metal hydride under a fixed temperature T.

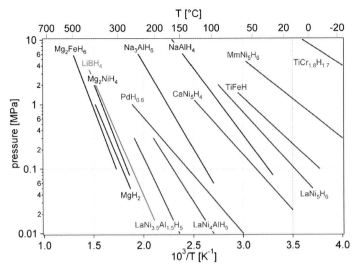

FIGURE 6.3 The van 't Hoff plots of several selected metal hydrides. *Source*: Reproduced with permission from Zuttel [3].

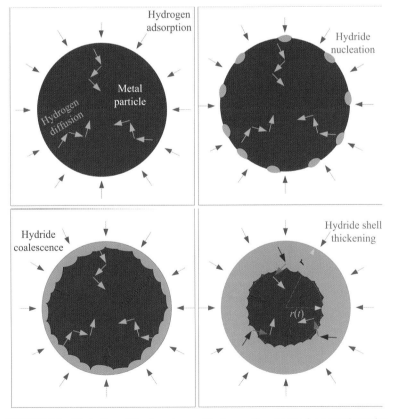

FIGURE 6.5 Illustration of the four stages of hydrogenation process for a metal powder.

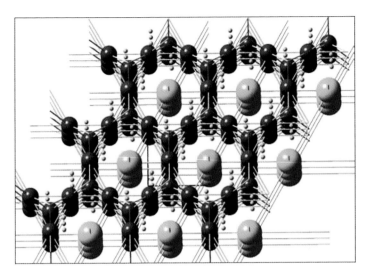

FIGURE 6.7 Structure of LaNi₅H₇. The sizes of the atom are declined from La, to Ni, to H. *Source*: Reproduced with permission from figure 2.52 in Sorensen [18].

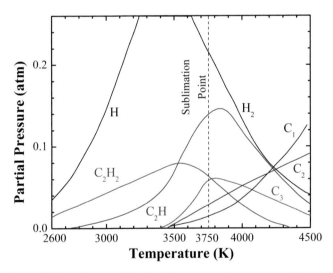

FIGURE 6.13 Equilibrium diagram of the C + H₂ system at 1 atm pressure. *Source*: Adapted with permission from Baddour and Iwasyk [39].

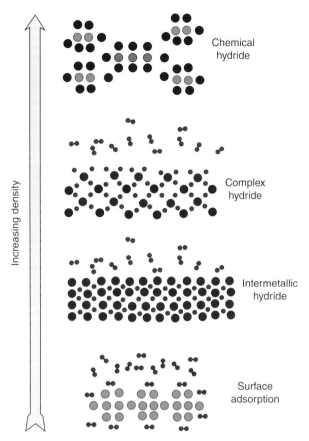

FIGURE 7.1 Hydrogen storage density in physisorbed materials, metal/complex, and chemical hydrides. *Source*: Reproduced with permission from Niemann et al. [1].

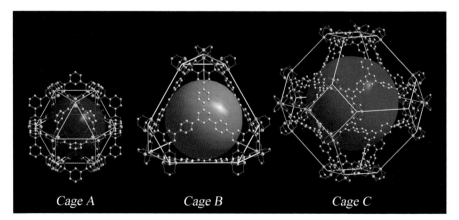

FIGURE 7.7 Different cages in the crystal structure of NOTT-112. Copper: blue–green; carbon: grey; oxygen: red. Water molecules and H atoms are omitted for clarity. *Source*: Reproduced with permission from Yan et al. [46].

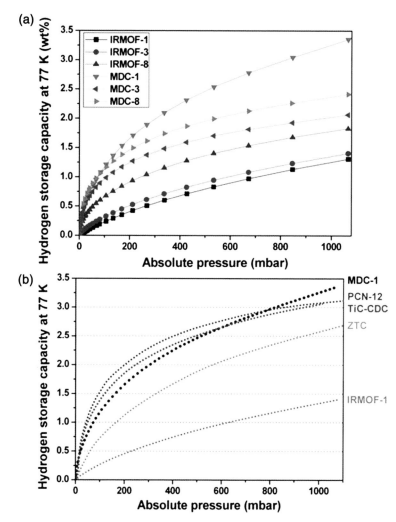

FIGURE 7.8 Hydrogen storage capacity at 77 K, and 1 bar of (a) the products and (b) the benchmark materials are shown for comparison. IRMOF stands for isoreticular MOF and MDC for MOD derived. *Source*: Reproduced with permission from Yang et al. [48].

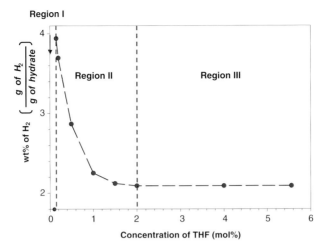

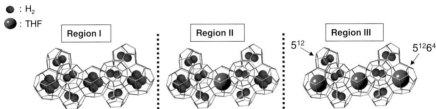

● : H₂
● : THF

FIGURE 7.9 H_2 gas content as a function of THF concentration, and a schematic diagram of H_2 distribution in the cages of THF+H_2 hydrate. (H_2 gas content is calculated from g of H_2 per g of hydrate, and expressed as wt%.) In region III, H_2 molecules are only stored in small cages, while in region II, both small and large cages can store H_2 molecules. At the highly dilute THF concentrations of region I, H_2 molecules can still be stored in both cages, but extreme pressures (~2 kbar) are required to form the hydrates. Pure H_2 clathrate $(2H_2)_2 \cdot (4H_2) \cdot 17H_2O$ would have a 5.002 wt% H2 content. *Source*: Reproduced with permission from Lee et al. [50].

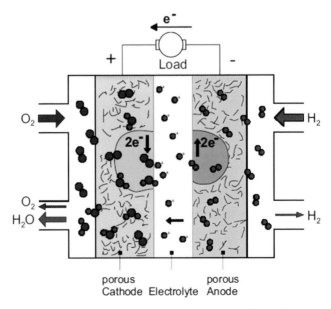

FIGURE 9.1 Schematic drawing of a hydrogen/oxygen fuel cell and its reactions based on the proton exchange membrane fuel cell. *Source*: Reproduced with permission from Carrette et al. [4].

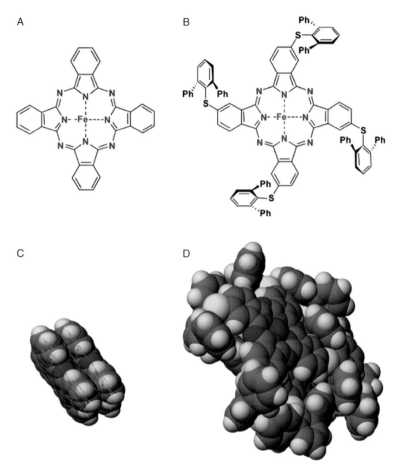

FIGURE 9.6 Atomic structure and the space filling stacking model of Fe–Pc (A,C) and Fe–SPc (B,D). *Source*: Reproduced with permission from Wu et al. [53].

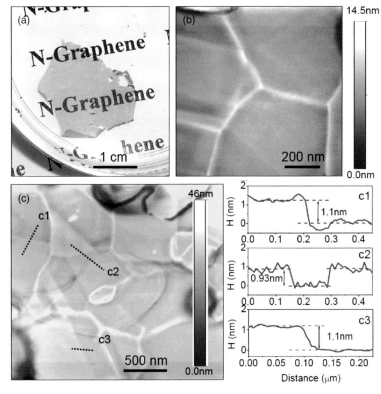

FIGURE 9.7 A digital photo image, AFM, and corresponding height analyses of the nitrogen doped grapheme. *Source*: Reproduced with permission from Gong et al. [54].

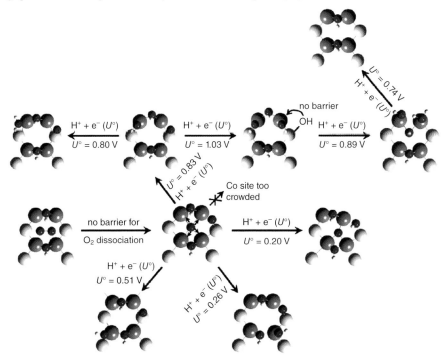

FIGURE 9.9 Top view of reduction pathways examined for the adsorbed O atoms on the partially OH(ads)-covered (202) surface of Co_9S_8. *Source*: Reproduced with permission from Sidik and Anderson [46].

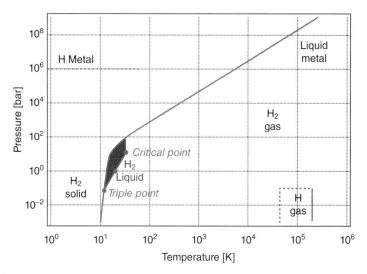

FIGURE 5.6   A simple phase diagram of hydrogen. *Source*: Reproduced with permission from Leung et al. [10].

The liquefaction process requires very clean hydrogen, several cycles of compression, liquid nitrogen or helium cooling, and expansion taking the advantage of the Joule–Thomson (JT) effect. When a real gas is allowed to expand adiabatically through a porous plug or a fine hole into a region of low pressure, it is accompanied by cooling (or heating). Cooling takes place because some work is done to overcome the intermolecular forces of attraction. In JT effect, the enthalpy H of the system remains a constant, and the JT coefficient is defined as

$$\mu = \left(\frac{\partial T}{\partial P} \right) H. \tag{5.20}$$

For cooling $\mu > 0$ and for heating $\mu < 0$. A unique locus of P–T points at $\mu = 0$ is called JT inversion curve (see Fig. 5.7, the maximum points of the T–P curve for a constant H). Thus, for a certain temperature, there exists a pressure beyond which $\mu < 0$ and isenthalpic expansion causes a temperature rise; while at lower pressures, $\mu > 0$ and isenthalpic expansion induces the cooling effect. Gases like H_2, He, whose inversion temperature is low, show heating effect at room temperature. However, if these gases are just cooled below the inversion temperature and then subjected to JT effect, they will also undergo cooling. For hydrogen, the maximum JT inversion temperature is at 205 K ($-68.15°C$). Thus, hydrogen needs to be precooled to below this temperature. This can be done using cold or liquid nitrogen. Then,

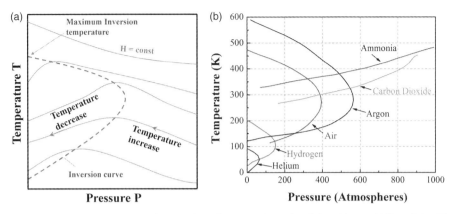

FIGURE 5.7 The illustration of the JT inversion curve (reproduced from Barron [11]) and the JT inversion curves for some conventional gases. *Source*: Reproduced with permission from Flynn [12]. (See color insert.)

the gas is further expanded to cool down to the boiling point of hydrogen using a JT valve or a cryogenic turbine.

The simplest liquefaction process is the Linde cycle, which is shown schematically in Figure 5.8 along with the *T–s* (entropy) diagram. Makeup gas is mixed with the uncondensed gas from the previous cycle, and the mixture at state 1 is compressed by an isothermal compressor to state 2. The high pressure gas is further cooled down after passing through a constant-pressure heat exchanger (ideally) by liquid nitrogen and the uncondensed gas from the previous cycle to state 3, and is then forced to pass through a throttle valve to state 4, which undergoes an adiabatic expansion, producing a saturated liquid–vapor mixture. The liquid (state 6) is collected as the desired product, and the vapor (state 5) is routed through the heat exchanger to cool the high pressure gas approaching the throttling valve. Finally, the gas is mixed with fresh makeup gas, and the cycle is repeated. A simple Linde cycle may not work for hydrogen at room temperature since its inversion temperature is very low, and physically it may be impossible to transfer enough energy in the heat exchanger to produce liquid. Therefore, modified Linde cycles with precooled gases may need, such as precooled Linde cycle, dual-pressure cycle, or even cascade Linde cycle [11].

The Claude cycle is a common method to liquefy high volume of hydrogen, as shown in Figure 5.9. The Claude cycle combines the isentropic (Brayton cycle) and isenthalpic (Linde cycle) expansions, and both the heat exchangers and mechanical expanders are used to cool the compressed and precooled hydrogen below its inversion temperature. As shown in Figure 5.9, the gas is first compressed, and passed through the first heat exchanger.

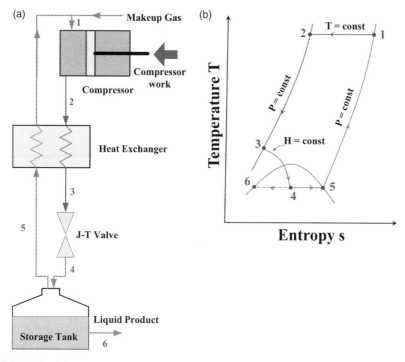

FIGURE 5.8 Schematic and temperature-entropy diagram of a simple Linde cycle. *Source*: Reproduced with permission from Barron [11]. (See color insert.)

Between 60% and 80% of the gas is then deviated from the mainstream, expanded through an expander. Such an expansion process is isentropic and a much lower temperature is attained than from an isenthalpic expansion. The portion of gas is reunited with the return stream below the second heat exchanger. The stream to be liquefied continues through the second heat exchanger, the third heat exchanger, and is finally expanded through a JT valve to the liquid tank. The cold vapor from the liquid tank is returned through the heat exchangers to cool the incoming gas. The Claude cycle may be used without modification to liquefy hydrogen since the system does not primarily depend on the expansion valve to produce low temperatures. In addition, by using liquid nitrogen precooling with the Claude system, a figure of merit 50–70% higher than that of the precooled Linde system may be obtained. Other methods, such as Haylandt cycle and dual-pressure Claude cycle, can be used to liquefy hydrogen also [11].

Liquid hydrogen needs to be stored in cryogenic vessels (or cryostats). The cryostats are metallic double-walled vessels with insulation, sandwiched between the walls. To minimize thermal losses, effects of thermal radiation,

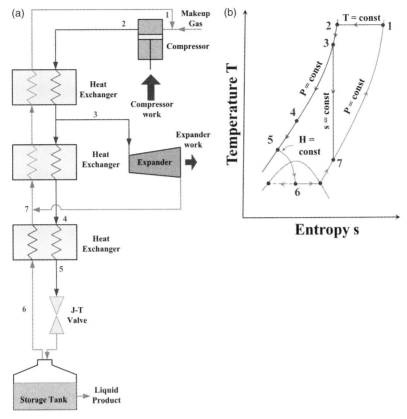

FIGURE 5.9   Schematic and temperature-entropy diagram of a simple Claude cycle. Reproduced with permission from Barron [11]. (See color insert.)

thermal convection, and thermal conduction have to be taken into account in designing the vessels (Fig. 5.10). The inner vessel is insulated with a multi-layered material with spacers acting as thermal barriers. The spacers are coated with high reflective Ag films to minimize thermal radiation loss. This inner vessel is mounted within the outer vessel by means of specially designed internal fixtures. The volume between the inner and outer vessels is evacuated to high vacuum to avoid possible heat leaks by thermal convection. In spite of the insulation, due to the unavoidable heat input, hydrogen will evaporate in the tank, which will cause the pressure rise in the vessel. Pressure build-up can be treated to be linearly proportional to storage time. Once the pressure reaches the maximum operation pressure of the tank, a blow-off valve has to be opened to release the hydrogen in order to maintain the safety of the system. The unexpected heat input could come externally or internally. As discussed in Chapter 1, hydrogen has two forms, the parahydrogen and orthohydrogen. These two forms of hydrogen not only have different internal

LH2-Tank System

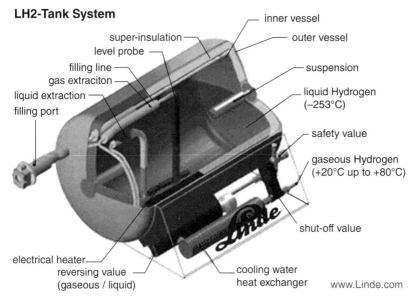

FIGURE 5.10 Schematic illustration of a representative cryogenic vessel. Source: "Hydrogen storage: state-of-the-art and future perspective," http://publications.jrc.ec.europa.eu/repository/bitstream/111111111/6013/1/EUR%2020995%20EN.pdf. (See color insert.)

energy, also have different thermal dynamic properties. The parahydrogen has lower melting and boiling points than those of the orthohydrogen. When hydrogen is cooled down, more orthohydrogen is converted to parahydrogen. The ratio of orthohydrogen can be reduced from 75% at room temperature to 25% at 77 K, and can be further reduced to 0.2% when the hydrogen is cooled down to the boiling point (20.08 K). The conversion of orthohydrogen to parahydrogen is a heat-release process. As long as there are orthohydrogens in the cryostats, conversion of orthohydrogen to parahydrogen is inevitable, which will cause heating of the liquid hydrogen. Clearly, although liquid hydrogen has significantly higher energy capacities compared with compressed hydrogen, it has some disadvantages, mainly the large amount of energy required to liquefy hydrogen, the strict requirements of cryogenic vessels with complicated thermal and pressure management, and hydrogen losses through evaporation from the containers.

5.4 SUMMARY

The compressed hydrogen tank and liquid hydrogen are the two most popular hydrogen storage methods for current industrial use. To use them as hydrogen vehicle, they will face similar challenges, such as vessel design and

material requirements, reducing energy expense in compression and liquefaction process, and therefore reducing the total cost. For the high pressure
gas storage, the challenge lies in increasing H_2 pressure, finding new light-
weight and higher strength materials for vessels. For liquid hydrogen storage,
the design of cryogenic vessel and reduction of the cost and energy loss in
liquefaction process present the main challenges. In addition, the concern of
safety of using pure hydrogen is an issue to the public. Thus, for future
hydrogen cars, safer and more reliable storage methods need to be developed,
and chemical storage is one alternative with high promise.

REFERENCES

1. Targets for onboard hydrogen storage systems for light-duty vehicles. http://www1
   .eere.energy.gov/hydrogenandfuelcells/storage/pdfs/targets_onboard_hydro_storage_
   explanation.pdf (last accessed December 9, 2013).

2. Status of hydrogen storage technologies. http://www1.eere.energy.gov/hydrogenand
   fuelcells/storage/tech_status.html (last accessed December 9, 2013).

3. Zhou, L., Zhou, Y.P. Determination of compressibility factor and fugacity coefficient of
   hydrogen in studies of adsorptive storage. *International Journal of Hydrogen Energy*
   2001, *26*(6), 597–601.

4. Wasserstoff Daten: Hydrogen data. http://www.h2data.de/ (last accessed December 9,
   2013).

5. Hydrogen Composite Tank Program. http://www1.eere.energy.gov/hydrogenandfuelcells/
   pdfs/32405b27.pdf (last accessed December 9, 2013).

6. Pathways to commercial success: Technologies and products supported by the hydrogen,
   Fuel Cells & Infrastructure Technologies Program. http://www1.eere.energy.gov/
   hydrogenandfuelcells/pdfs/pathways_success_hfcit.pdf (last accessed December 18,
   2013).

7. Silbey, R.J., Alberty, R.A., Bawendi, M.G. *Physical Chemistry*, 4th ed., John Wiley &
   Sons, Hoboken, NJ, 2005.

8. Strobel, R., Oszcipok, M., Fasil, M., Rohland, B., Jorissen, L., Garche, J. The compression of hydrogen in an electrochemical cell based on a PE fuel cell design. *Journal of
   Power Sources* **2002**, *105*(2), 208–215.

9. Abdulla, A., Laney, K., Padilla, M., Sundaresan, S., Benziger, J. Efficiency of hydrogen
   recovery from reformate with a polymer electrolyte hydrogen pump. *AIChE Journal*
   2011, *57*, 1767–1779.

10. Leung, W.B., March, N.H., Motz, H. Primitive phase diagram for hydrogen. *Phys. Lett.
    A* **1976**, *56*, 425.

11. Barron, R.F. *Cryogenic Systems*, 2nd ed., Oxford University Press, New York, 1985.

12. Flynn, T.M. *Cryogenic Engineering*, Dekker, New York, 1997.

6

Chemical Storage Based on Metal Hydrides and Hydrocarbons

6.1 BASICS ON HYDROGEN STORAGE OF METAL HYDRIDES

Many metals, intermetallic compounds, and alloys can react with hydrogen to form solid metal hydrides. Most metal hydrides have hydrogen very strongly bound to the metal. Some of the metal hydrides have even higher volumetric hydrogen storage capability than that of liquid hydrogen, which makes the metal hydride very attractive for on-board hydrogen storage applications. For example, Mg_2FeH_6 and $Al(BH_4)_3$ have the highest volumetric hydrogen density known today, 150 $kg m^{-3}$.

Figure 6.1 shows a comparison of both the volumetric and gravimetric storage capacities of different materials summarized by Züttel recently [1]. In general, most metal hydrides have very high volumetric density compared with that of liquid hydrogen, but the gravimetric density is usually low. Also, metal hydrides have the advantage for low pressure hydrogen adsorption and storage, thus making them more attractive from safety point of view. Since metal hydrides are in solid state, they do not require a complicated container to store, which makes them even more attractive. However, the formation of metal hydride and the dehydrogenation process are both chemical processes, which involve the formation or breaking of metal–hydrogen bonds, and the hydrogen atoms often occupy interstitial sites of metal. As a result, relatively high temperatures, around 120–300°C, are

Hydrogen Generation, Storage, and Utilization, First Edition. Jin Zhong Zhang, Jinghong Li, Yat Li, and Yiping Zhao.
© 2014 John Wiley & Sons, Inc. Published 2014 by John Wiley & Sons, Inc.

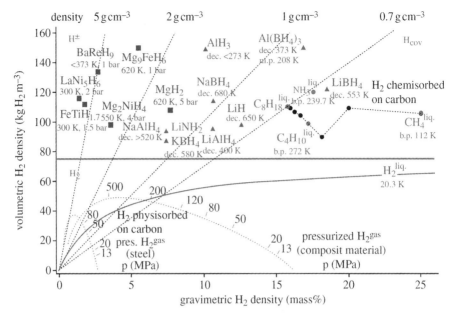

FIGURE 6.1   Volumetric and gravimetric hydrogen density of some selected hydrides. *Source*: Reproduced with permission from Züttel et al. [1].

required to release the hydrogen, which presents a major challenge for metal hydrides to meet the target for onboard hydrogen fuel systems, that is, <100°C for hydrogen release and <700 bar for hydrogen recharge (20–60 kJ·mol^{-1} H$_2$). In addition, all the metal hydrides that can operate at ambient temperature and pressure consist of transition metals and have a relatively low gravimetric density (usually <10 wt%). Therefore, exploring lightweight metal hydride and tuning their thermodynamic and kinetic properties are of strong interest.

6.2 HYDROGEN STORAGE CHARACTERISTICS OF METAL HYDRIDES

Since the hydrogen storage mechanisms of metal hydrides and complex hydrides involve primarily chemical reaction processes, their storage characteristics are more complicated than that of physical storage of pure hydrogen discussed in Chapter 5. To determine whether the material has good hydrogen storage performance, the following properties need to be examined.

6.2.1 Storage Capacities

Both the gravimetric and volumetric storage capacities are still the most important parameters to compare when measuring the storage performance of different materials. The ideal storage capacity of a metal hydride is determined by the stoichiometry of the particular hydride. For a metal hydride MH_x, the ideal gravimetric storage capacity ρ'_M is determined as

$$\rho'_M = \left[\frac{xM_H}{xM_H + M_M} \times 100 \right] \text{wt}\%, \tag{6.1}$$

where M_H and M_M are the atomic mass of hydrogen and metal (or alloy) M. The volumetric storage capacity ρ'_V is defined as

$$\rho'_V = \frac{m_H}{V_M} = \frac{xM_H}{M_M / \rho_M}, \tag{6.2}$$

where m_H is the mass of hydrogen stored in the metal with a volume V_M, and ρ_M is the mass density of metal M. Here, the definition does not consider the lattice expansion during hydrogenation of a metal. A more rigorous definition that can be applied for any chemical storage materials is,

$$\rho'_V = \frac{m_H}{V_{MH_x}} = \frac{xM_H}{M_{MH_x} / \rho_{MH_x}}. \tag{6.3}$$

In practice, the hydrogen storage capacity may vary since there may be impurities or defects in the materials, and hydrogen may adsorb into the material through physical interactions. Furthermore, the thermodynamics and kinetics of hydrogenation and dehydrogenation processes will determine the real hydrogen storage capacity under specific conditions.

6.2.2 Thermodynamics and Reversible Storage Capacity

The hydrogenation and dehydrogenation are the mutual reverse processes for a metal hydride and only occur at a certain temperature (preferably at room temperature) and pressure range, that is, depending on the thermodynamic nature of the metal and metal hydride. When exposed to hydrogen, the metal or metal alloy (M) will form metal hydride through the following reaction,

$$M + \frac{x}{2}H_2 \rightleftharpoons MH_x + Q, \tag{6.4}$$

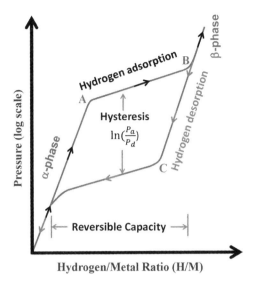

FIGURE 6.2   The typical PCT curves for the hydrogenation and dehydrogenation of a metal hydride under a fixed temperature T. (See color insert.)

where Q is the heat of hydride formation. In general, the hydrogenation is exothermic, while the dehydrogenation is endothermic. Practically, the hydrogenation process happens at high hydrogen pressure, while dehydrogenation occurs at low pressure. Figure 6.2 shows a typical pressure–composition–temperature (P–C–T) curve during a hydrogenation and dehydrogenation cycle. With the increasing hydrogen pressure, a metal starts to adsorb hydrogen to form metal–hydrogen solid solution (α-phase). When the pressure reaches "A" location shown in Figure 6.2, the metal starts to form hydride (β-phase). At this stage, the hydrogen pressure (P_A) almost remains as a constant while the hydrogen content increases significantly. The hydrogenation process will be complete at "B" location. This A–B adsorption plateau characterizes the effective hydrogen storage capacity at a fixed temperature. In general, the adsorption plateau pressure will increase with temperature, and follow the van 't Hoff relation [2],

$$\ln P = \frac{\Delta H}{RT} - \frac{\Delta S}{R}, \tag{6.5}$$

where P is the hydrogen pressure, ΔH and ΔS are the enthalpy and entropy of hydride formation or decomposition, R is the universal gas constant, and T is the temperature. The heat of formation can be obtained by plotting the plateau pressure $\ln P$ versus $1/T$ (van 't Hoff plot), as shown in Figure 6.3

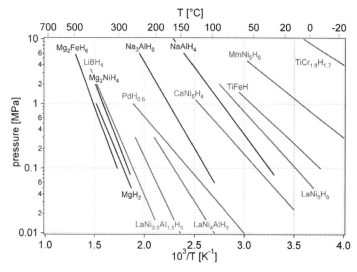

FIGURE 6.3   The van 't Hoff plots of several selected metal hydrides. *Source*: Reproduced with permission from Zuttel [3]. (See color insert.)

[3]. The dehydrogenation process is a reverse process as shown in Figure 6.2, and it also shows a desorption plateau with a lower near-constant hydrogen pressure P_D. The adsorption and desorption plateaus form a hysteresis loop for the hydrogenation and dehydrogenation processes, and the free energy difference associated with the hysteresis is given by

$$\Delta G_{\text{hyst}} = RT \ln\left(\frac{P_A}{P_D}\right).$$
(6.6)

The hysteresis represents a loss in the efficiency of the hydride due to irreversible degradations of materials during hydrogenation/dehydrogenation processes. A good hydride material should have a hysteresis as small as possible.

The width of the plateau, $\Delta(H/M)_r$, as indicated in Figure 6.2, is defined as the reversible capacity, which is usually smaller than the maximum (ideal) storage capacity discussed earlier. This is an important parameter for practical applications.

6.2.3 Hydrogenation and Dehydrogenation Kinetics

The rates of hydrogen adsorption and desorption are two very important parameters for characterizing metal hydrides. These rates depend on the

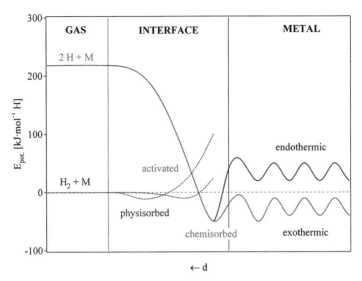

FIGURE 6.4 Potential energy curve for hydrogen binding to a metal: physisorption for both activated and nonactivated processes; dissociation and surface chemisorption; surface penetration and chemisorption on subsurface sites; and diffusion. *Source*: Reproduced with permission from Zuttel [3].

detailed hydrogen sorption kinetics, the sorption temperature, and pressure. Microscopically, this hydrogenation process involves several steps that can be described by Lennard-Jones potential as shown in Figure 6.4 [4]. When a hydrogen molecule approaches the metal surface, it encounters successive potential minima corresponding to molecular adsorption, atomic adsorption, and bulk absorption. It is first physisorbed on the surface of the metal due to van der Waals forces or electrostatic attraction. Such a weak interaction usually prevents significant absorption at room temperature. At high enough pressure and temperature, the adsorbed hydrogen molecule can be dissociated at the surface by transferring an electron between the metal and the hydrogen and becomes chemisorbed. This step may require thermal or catalytic activation due to the dissociation energy barrier. After surface chemisorption, the hydrogen atoms move to subsurface sites, rapidly diffuse through the material, and become a solution of H in the metal (α-phase, see Fig. 6.2). As the hydrogen concentration in the α-phase increases, a more stable metal hydride phase (β-phase) is formed. Such a phase transition is usually characterized by a crystalline structure change, a volume expansion, and a nucleation energy barrier associated with volume expansion and interface energy between the phases. Macroscopically, the detailed hydrogenation process can be divided in four different stages as shown in Figure 6.5

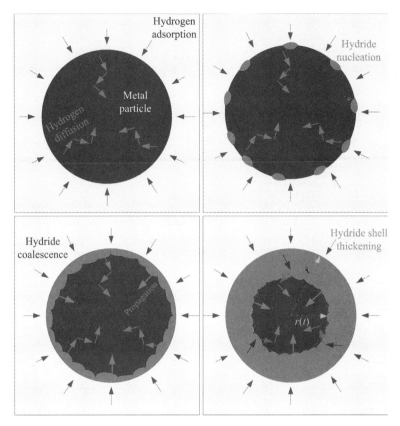

FIGURE 6.5   Illustration of the four stages of hydrogenation process for a metal powder. (See color insert.)

[5]. Initially, the hydrogen will adsorb on the metal surface and diffuse into the metal. At appropriate temperature, hydrogen starts to react with the metal to form metal hydride. The reaction starts at the surface of the metal, and propagates into the metal core. At the beginning, random hydride patches will be formed on the metal surface, similar to a nucleation process. With the progress of the reaction, the hydride patches will coalescence and form a hydride shell around a metal particle, and subsequent reactions will involve propagation of this hydride shell. The formation of the hydride shell also has another negative effect, that is, to impede the hydrogen diffusion from the atmosphere into the inner core of the particle (the hydrogen diffusion rate in metal is in general much faster than that in metal hydride). Therefore, multiple rate limit processes could determine the hydrogenation kinetics.

The kinetics of hydrogenation of metal hydride powder can be modeled using a single particle analysis (SPA) method based on the hydride shell propagation by neglecting the initial nucleation process (also ignoring the self-heating effect and volume expansion during the hydrogenation process). For simplicity, we only consider a spherical particle with radius R. If the reaction rate of the hydride shell is controlled by hydride formation, the shell will propagate with a constant velocity v, and the radius of metal core will shrink as, $r(t) = r - vt$. Thus, the volume fraction α of hydride in the particle shall follow [5]:

$$\alpha = 1 - \left(1 - \frac{v}{r}t\right)^3.$$ (6.7)

For a hydrogen diffusion-controlled process, α is determined by [5]:

$$1 - \frac{2}{3}\alpha - (1-\alpha)^{\frac{2}{3}} = 2k_D t / r^2,$$ (6.8)

where k_D is a parameter associated with hydrogen diffusion. Considering that there are different-sized (also shaped) powders in the real sample, one needs to take the particle size distribution $f(r)$ into consideration. Thus, at time t, the total volume of metal hydride V_r is given by [5]:

$$V_r(t) = \int_{r_{\min}}^{\infty} \frac{4}{3}\pi r^3 \alpha(r, t) f(r) dr,$$ (6.9)

Therefore, the hydrogen content $\rho'_H(t)$ can be expressed as:

$$\rho'_H(t) = \frac{V_r(t)}{V_0} \frac{xM_H}{xM_H + M_M},$$ (6.10)

where

$$V_0 = \int_{r_{\min}}^{\infty} \frac{4}{3}\pi r^3 f(r) dr,$$

is the total volume of the powder, and $r_{\min}$ is the minimum radius of the powder particles.

For most practical cases, the hydrogenation process can be fit by the Avrami's model [6],

$$\alpha = 1 - \exp(-kt^n),$$ (6.11)

where k is treated as the apparent reaction constant and n depends on the geometry and dimensionality of the reaction. The reaction rate k is thermally activated and follows the Arrhenius relation,

$$k = k_0 \exp\left(-\frac{E_a}{k_B T}\right), \tag{6.12}$$

where E_a is the reaction activation energy and k_B is the Boltzmann constant. The activation energy for adsorption and desorption may be different for the same metal hydride, depending on the structure, morphology, and catalyst used. To lower the activation energy, different catalysts are usually added to the metal hydride systems.

6.2.4 Cycling Stability

For practical applications, the ability of the metal hydride to retain its reversible storage capacity during repeated hydrogenation and dehydrogenation cycles is an important parameter to consider. Different applications may require different cycling times. For example, the 2015 DOE storage target specifies 1500 cycles for a good hydrogen storage material [7]. There are intrinsic and extrinsic degradations governing the cycling stability of a metal hydride. The intrinsic degradation is caused by the physical and chemical changes of the metal hydrides during the cycling process. The degradation includes hydride decomposition into more thermally stable products, formation of defects and dislocations, loss of reversible storage capacity, and loss of material components during the cycling process. In particular, the decrepitation, a self-pulverization of metal hydrides into smaller powder due to volume change during cycling, not only will change the heat transfer and gas flow inside the storage tank, but also will induce tank rupture and affect the purity of the released hydrogen. These intrinsic degradation processes are irreversible. The extrinsic degradation is due to impurities in the H_2 gas during the hydrogenation process. Gas impurities, such as O_2, CO, CO_2, SO_2, H_2O, H_2S, NH_3, hydrocarbons, formaldehyde, and formic acid, could react with and poison or corrode the metal or metal hydride surfaces, reducing hydrogen adsorption/desorption. If the hydrides are poisoned, the material could be regenerated by using high purity H_2 cleaning and high temperature annealing.

6.2.5 Activation

Usually, the surfaces of metals are covered by a layer of oxide, or adsorbed with a layer of gas or water vapor, which makes them hard to be hydrogenated. Such a hydrogen barrier layer must be broken or removed in order to accelerate the hydrogenation process. Thus, initial hydrogenation could be performed at high temperature and high hydrogen pressure in order to quickly reduce the surface oxide and remove surface adsorbents. The dehydrogenation will then proceed. Such a cycling process will be continued for several times until the hydrogenation/dehydrogenation processes become stable. Such an activation process is needed for most metal hydrides.

6.3 DIFFERENT METAL HYDRIDES

The details of different metal hydrides and their properties can be found at U.S. DOE Database http://hydrogenmaterialssearch.govtools.us/. There are some very good recent books focusing on hydrogen storage materials by Broom (2011) [8], Hirscher (2010) [9], and Zuttel (2008) [10]. In the following, we briefly discuss some most promising metal hydride materials.

6.3.1 Binary Metal Hydrides

Binary hydrides contain only a single element of metal. They are generally either too unstable or too stable for use as practical storage materials. The most famous example for binary hydride is magnesium hydride, MgH_2. It has attracted by far the most attention for binary hydrides due to its relatively high gravimetric density, 7.6 wt%, and it has already been demonstrated for onboard applications [11–13]. However, for bulk MgH_2, there are two key drawbacks keeping it from real onboard applications: the hydrogen desorption temperature for MgH_2 being high, at 330°C, which is due to its high formation enthalpy, $-74.5 \cdot kJ\ mol^{-1}\ H_2$, and the slow kinetics near the ambient conditions [11, 14]. To improve the performance of MgH_2, many different processes have been suggested. One particular method is to reduce the size of MgH_2 to nanometer scale through ball milling, and another way is to add catalysts. Figure 6.6 shows the desorption kinetics of MgH_2 powder with 0.1 mol% Nb_2O_5 catalyst after ball milling for different time durations [15]. The longer the milling time, the faster the desorption kinetics. Other notable binary hydrides include AlH_3, LiH, and PdH_x. AlH_3 has a very high gravimetric density, ∼10.1 wt%, but the hydrogen desorption requires very high temperature. PdH_x can adsorb and desorb hydrogen at relatively low temperature, but it is expensive and has low gravimetric density (0.6 wt%).

6.3.2 Metal Alloy Hydrides

When two or more metallic elements are combined and react with hydrogen, they form the ternary system AB_xH_y. Usually, the element A is a rare earth

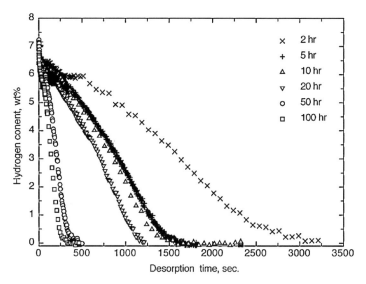

FIGURE 6.6   Desorption curves of MgH$_2$ at 573 K in vacuum with Nb$_2$O$_5$ catalyst and milled for 2, 5, 10, 20, 50, and 100 hours. *Source*: Reproduced with permission from Barkhordarian et al. [15].

or an alkaline metal and can form a stable hydride, while element B is a transition metal and forms an unstable hydride. The resultant alloy or inter-metallic compound tends to form a hydride of intermediate stability. Some well-known intermetallic compounds are with $x = 0.5, 1, 2$, and 5. Most intermetallics have low gravimetric density, and some require high sorption temperature. Therefore, they are not yet suitable for onboard applications. A well-studied AB$_5$ intermetallic is LaNi$_5$, which can form a hydride (LaNi$_5$H$_7$) under moderate hydrogen pressures and ambient temperatures (see Fig. 6.7) [16]. Its enthalpy of formation is -15.7 kJ·(mol·H)$^{-1}$ and enthalpy of decom-position is -15.1 kJ·(mol·H)$^{-1}$ [8]. Its reversible gravimetric density is 1.25 wt% [17]. For all the AB$_5$ intermetallics, the gravimetric storage capac-ity is substantially lower than the current U.S. DOE target for mobile hydro-gen storage applications. However, the AB$_5$ intermetallics have some remarkable cycling performance and high volumetric storage density, and are therefore a prime example of practically effective reversible hydrogen storage materials.

6.3.3 Complex Metal Hydrides

Complex hydrides have the highest hydrogen storage density among all the metal hydrides. They are salt-like materials in which hydrogen is either ioni-cally or covalently bound to the storage material. Complex hydrides have the

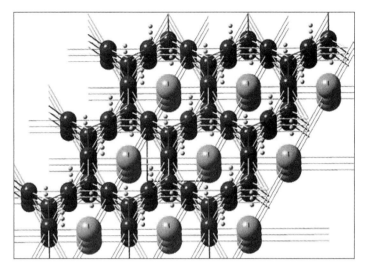

FIGURE 6.7 Structure of $LaNi_5H_7$. The sizes of the atom are declined from La, to Ni, to H. *Source*: Reproduced with permission from figure 2.52 in Sorensen [18]. (See color insert.)

chemical formula of $A_xB_yH_z$, where A is the element in the first or second group of the periodic table, and element B is either aluminum, nitrogen, or boron. The corresponding complex hydrides are called alanates, nitrides, and borohydrides, respectively. These materials are recently under intensive investigation due to the discoveries of their reversible sorption capability [19, 20]. One specific example is sodium alanate ($NaAlH_4$), the dehydrogenation of whcih involves two related reactions,

$$3NaAlH_4 \rightarrow Na_3AlH_6 + 2Al + 3H_2 \text{ at } 210 - 220°C \qquad (6.13)$$

$$Na_3AlH_6 \rightarrow 3NaH + Al + 1.5\,H_2 \text{ at } 250°C. \qquad (6.14)$$

This process is irreversible and the desorption kinetics is also slow, which makes the material questionable for storage application. However, the discovery of the reversible sorption of $NaAlH_4$ by Bogdanovic and Schwickardi in 1997 by adding $TiCl_3$ into the structure has changed the view on complex metal hydrides.[19] Ti-doped $NaAlH_4$ can desorb hydrogen at 120°C, be rehydrogenated at 170°C, and has gravimetric density around 7.5 wt% [21]. The material has already been used in practical hydrogen storage units.

6.3.4 Improving Metal Hydride Performance

Although there are many advantages with using metal hydrides for onboard vehicle applications, two major problems prevent them from practical

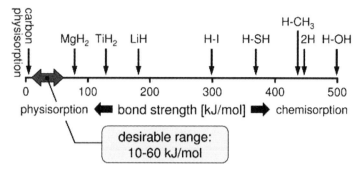

FIGURE 6.8   Targeted range of bond strengths that allow hydrogen release around room temperature. A given material can exhibit both chemisorption and physisorption. *Source*: Reproduced with permission from Berube et al. [22].

application: high adsorption/desorption temperature and slow kinetics. The strong chemical bonds (>50 kJ mol^{-1}) formed between hydrogen and metals in metal hydrides result in the high storage capacity and stability at room temperature, and also lead to an enormous energy release during the hydrogenation process. However, to release hydrogen, the hydride must be heated to high temperature in order to supply sufficient energy to break the strong chemical bonds. Thus, to obtain a destabilized hydride with small binding energy that is in thermodynamic equilibrium with the hydrogen gas closer to room temperature is the key to reducing the heat of formation which would in turn reduce the release temperature [22]. Figure 6.8 shows the range of binding energies E_b targeted by DOE. For metal hydrides, the binding energies are too high for chemisorption, while for carbon structures, metal organic frameworks, and so on, the binding energies are too low due to physisorption (<10 kJ·mol^{-1}). Therefore it is necessary to raise E_b for physisorbed materials, or lower E_b for metal hydrides through thermodynamic destabilization. Since storage capacity remains a priority, it is important to achieve destabilization while minimizing the storage capacity reduction. Many different methods or processes have been developed to destabilize metal hydrides. Among them, to make metal hydride as nanometer-size crystals and to add nano-sized catalysts into metal hydrides have attracted the main attentions and also led to significant improvement for metal hydride performance.

Previous studies on nanocrystalline metal hydrides have demonstrated the following potential advantages for hydrogen storage:

1. *Nanostructures Will Improve the Hydrogen Adsorption and Desorption Kinetics.* Numerous experiments indicated that for all metal hydrides the kinetics of both adsorption and desorption can be improved by an

order of magnitude simply by reducing the grain size of the metal [23–26]. For example, while it is very hard to activate and hydrogenate polycrystalline Mg films at 300°C, when the grain size is reduced to 30–50 nm, the Mg films absorb hydrogen more quickly [23–26]. Research performed on thin films of pure magnesium demonstrated that the thinner the magnesium sheet is, the faster it achieves complete formation of MgH_2 [27–29]. Similar results have been observed for other materials, such as Mg_2Ni, FeTi, and $LaNi_5$, as well [23–26]. The crystalline structure of the resulting materials has been found to play a significant role. For instance, crystalline, amorphous, nanocrystalline, or the mixture of amorphous and nanocrystalline phases affect the thermodynamics of hydrogen adsorption for FeTi. In some cases, a partially amorphized alloy may result in a further increment of the hydrogen solubility [23–26]. It has been suggested that the grain boundaries play a critical role in improving the hydrogenation properties.

2. *Nanostructures Can Lower the Desorption Activation Energy.* For example, theoretical calculations based on quantum chemistry have shown that small MgH_2 clusters have much lower desorption energy than bulk MgH_2 [30]. The hydrogen desorption energy decreases significantly when the crystal grain size becomes smaller than 1.3 nm.

3. *Nanoscaled Catalyst Can Greatly Improve Hydrogen Adsorption.* Effective catalysts, even added in small amounts, enhance the formation of a hydride to a reasonable extent. Small nanoscale catalysts distributed over the surface of nanocrystalline hydrides can result in a spectacular improvement of sorption properties, such as elimination of the need for activation and overall improvement of sorption kinetics [23–26]. Previous experiments have shown that with proper catalysts, the activation time for $LaNi_5$ nanocrystalline film to absorb hydrogen can be eliminated [25]. Therefore, by eliminating the need for activation, the combination of nanostructure and nanocatalysis yields a tremendous gain in sorption kinetics.

4. *Nanocomposites Can Further Improve the Quality of Hydrogen Adsorption.* A specifically designed nanocomposite, where both nanostructure and nanocatalyst can be incorporated into a more complex system, can have properties surpass that of the individual components alone. For example, a composite of a nano-mixture of a high temperature hydride (Mg) with a low temperature hydride (FeTi, or $LaNi_5$) can be used for "cold-starting." [25]. Or the "cascade" of components can be operated at various temperatures and hydrogen pressures. Doping of

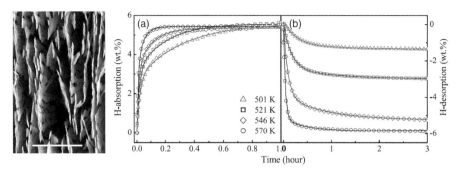

FIGURE 6.9   (Left) SEM image of the Mg nanoblades, and (right) the sorption curves of (a) hydrogen absorption under a hydrogen pressure of 10 bar and (b) hydrogen desorption under vacuum at varying temperatures for V decorated Mg nanoblade array. *Source*: Reproduced with permission from He et al. [31].

nanostructures with metal ions such as $Ti^{3+/4+}$ is another useful method to improve the hydrogen storage properties of the nanostructures [25].

As an example, Figure 6.9 shows a 4.6 at% vanadium decorated Mg nanoblade structure made by oblique angle deposition [31]. The blade thickness is about 65 nm, and the height is about 50 μm. The hydrogen sorption kinetic curves at various temperatures ($T \leq 570$ K) are also plotted. The reversible hydrogen amount is less than 6 wt%, which is slightly lower than the theoretical value 7.6 wt% of MgH_2. The V-decorated Mg nanoblades can absorb hydrogen to saturation within 7 minutes at 570 K with a rate constant of $k \approx 131$ h^{-1}. The hydrogenated nanoblades can desorb hydrogen almost completely within 15 minutes at 570 K with $k \approx 26$ h^{-1}. By plotting the obtained rate constant k versus reciprocal temperature $1/T$ in Figure 6.10, the activation energy is estimated to be $E_a^a = 35.0 \pm 1.2$ kJ $\cdot (mol \cdot H_2)^{-1}$ for absorption process and $E_a^d = 65.0 \pm 0.3$ kJ for desorption process. These activation energies are much lower than the desorption energy of 141 kJ $\cdot (mol \cdot H_2)^{-1}$ for MgH_2 film [32] and 156 kJ $\cdot (mol \cdot H_2)^{-1}$ for MgH_2 powder [33], indicating the catalytic effect of the V coating on the MgH_2 formation and decomposition.

However, the fundamental understanding of how those nanostructures improve the thermodynamics and kinetics of metal hydrides is still under debate, although some consensuses have been reached. In the past 5 years, a number of research groups have conducted extensive experimental and theoretical research toward such an understanding. We summarize the main results and challenges from those studies following the arguments by the MIT groups [22, 34]:

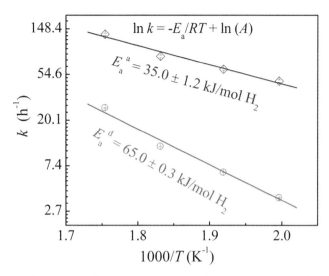

FIGURE 6.10   Arrhenius plots of hydrogen absorption and desorption rate constant k versus reciprocal temperature $1/T$ for V-decorated Mg nanoblades. *Source*: Reproduced with permission from He et al [31].

1. *The Effect of Nanostructure on the Thermodynamics of Hydriding/ Dehydriding of Metal Hydrides.* Four potential factors could be account to affect the thermodynamics of hydrogen storage process: the presence of a metastable phase, nanostructure size (surface area), stress/strain at the grain boundary, and excess volume in deform regions.
   (a) New chemical species can be introduced to react with metal and form an intermediate state. This can alter the formation path of metal hydride, and potentially reduce the heat of formation [22, 35]. Strictly speaking, this effect is not limited to nanostructures.
   (b) The size of the metal nanostructure could also have an effect on lowering the heat of formation [22]. Smaller size means more surface energy. If the surface energy of the hydride layer is larger than that of the metal, due to the extra energy deposited on the surface, the heat of formation will be reduced [22, 36]. Calculations show that in order to achieve a significant reduction of the heat of formation, for Mg, the nanoparticle's diameter should be smaller than 2 nm [22, 36].
   (c) The grain boundaries caused by mismatched crystal plane orientation could also supply extra energy for lowering the heat of formation [22]. According to a conservative estimate, for grain size on the order of 7–9 nm, the maximum reduction in the heat of

formation is about 9–10 kJ·mol^{-1} [35]. This reduction cannot explain large enthalpy changes observed in experiments.

(d) The most probable enthalpy reduction cause may be the excess volume effect. In heavily milled metal hydride samples, it is likely that noncrystalline regions arise and the material is deformed. These deformations could be gathered around grains or particles surfaces. The resulting lattice distortions will change the energy level of the metal and hydride and therefore could explain the formation enthalpy change [35].

2. *The Effect of Nanostructure on the Kinetics of Metal Hydriding/ Dehydriding.* The kinetic process of hydrogen storage comprises several steps: the dissociation and penetration of hydrogen at the interface, the formation of metal hydride with moving boundary, the diffusion of hydrogen, along with the heat dissipation and stress/strain change. As the hydrogenation reaction progresses, the rate limiting process changes from the dissociation and penetration of hydrogen at the interface to the nucleation of the β-phase, and finally to the diffusion of hydrogen through the β-phase layer formed around the particle (see Figure 6.5). For the desorption process, the main rate limiting processes are the slow diffusion through the β-phase layer and the high hydrogen dissociation energy barrier. The absorption kinetics is accelerated by the high reaction rate, the large diffusion coefficient, and the small diffusion length, that is, small particle size. By engineering metal hydrides into nanostructures, the increased surface area and porosity of nanostructures can offer a larger number of dissociation sites and allow fast gaseous diffusion to the center of the material [22]. If the nanostructures accompany with increased volume of grain boundaries, those grain boundaries will weaken the binding between metal and hydrogen atoms, which helps the site-to-site hopping for hydrogen and enhances diffusion in the α-phase. Thus, as nanostructures favor the thermodynamics of hydrogen absorption, they also improve the kinetics. Several experiments have already demonstrated that the hydrogen diffusivity increases with decreasing particle size [37, 38]. The grain boundaries and internal strains in nanostructures also promote a fast kinetics [22]. In addition, the kinetics can be dramatically improved by a process known as spillover through the use of a proper surface catalyst on the metal surface as shown in Figure 6.11 [22]. In the spillover process, the hydrogen molecule is dissociated on the metal catalyst, and the resulting hydrogen atom diffuses to the surrounding storage media. Such a process can also make the diffusion through the surface insensitive to the oxide layer, which prevents the need for an activation

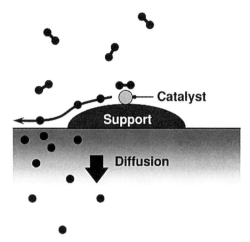

FIGURE 6.11 The spillover mechanism: the hydrogen molecules dissociate on the catalyst. Some hydrogen atoms remain attached to the catalyst, while others diffuse to the catalyst support and subsequently penetrate into the metal, where the hydrogen is said to spill over and interact directly with the metal. *Source*: Reproduced with permission from Berube et al. [22].

process and increases the resistance to contaminants and air exposure. In addition, as shown in Figure 6.5, the later kinetics of the hydrogenation process is governed by the diffusion of hydrogen atom through the β-phase layer formed around the particle. Since the hydrogen diffusion rate in the β-phase can be significantly smaller than that in the α-phase, minimizing this effect could promote faster kinetics. Figure 6.12 shows that for a sufficiently small particle, a closed β-phase layer may not be formed, so that the hydrogen atoms could have fast pathways to access the particle core during the hydrogenation reaction or leave the core upon the dehydrogenation reaction [22]

6.4 HYDROCARBONS FOR HYDROGEN STORAGE

As discussed in Chapter 2, hydrocarbons are organic compounds that contain hydrogen and carbon atoms. In principle, hydrocarbons could be used for hydrogen storage. In practice, this is quite challenging, since hydrocarbons are generally very stable under ambient conditions, and the release of hydrogen from hydrocarbons is a highly endothermic process. High temperature steam reforming is a common process for hydrogen generation from hydrocarbons, as discussed in detail in Chapter 2.

Upon hydrogen release, either carbon (C) or carbon dioxide (CO_2) is produced as a by-product. For the purpose of hydrogen storage, it would be

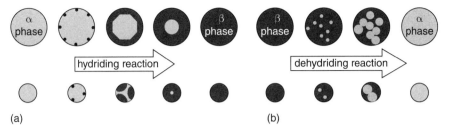

FIGURE 6.12 Comparison of the hydrogenation and dehydrogenation of a large and a small particle: (a) for hydrogenation in a large particle (upper raw), multiple nucleation sites at the surface will merge and form a closed layer that prevents fast diffusion of hydrogen to the core of the particle, slowing down the kinetics of the α- to β-phase transition considerably. If the particle is small (lower raw), fast diffusion of the hydrogen through the α-phase remains possible for a larger fraction of the α- to β-phase transition and (b) for desorption in a large particle (upper raw), hydrogen has to diffuse through a thicker layer of the β-phase before being released, while hydrogen rapidly reaches the surface for a smaller particle (lower raw). *Source*: Reproduced with permission from Berube et al. [22].

necessary to hydrogenate carbon or CO_2 back to hydrocarbons if a sustainable and recyclable system is needed. The reaction of hydrogen with either carbon or CO_2 is complex and requires special conditions, such as high pressure and high temperature or assistance with catalysis. In the next section, we will discuss the two reactions separately.

6.4.1 Reaction between Carbon Atom and Hydrogen

This seemingly simple reaction is in reality quite complex. Interestingly, the study of the reaction between C and H or H_2 is very limited to date, partly due to the need to produce C and H in a well-controlled environment. One of the earlier studies found that the main products of the reaction between carbon vaporized in a high intensity arc reactor operating at 2800 K and hydrogen passing through the arc were acetylene, hydrogen, and condensed carbon when the hot reaction mixture was sampled under fast-quenching conditions [39]. Acetylene contents as high as 18.6 V% at 1 atm in the quenched gas were obtained without diluent, and as high as 23.8 V% with 63.6% helium diluent. It was suggested that the C_2H radicals existed in the hot gas mixture in appreciable concentrations, which were generated from dissociation of C_2H_2 resulting from the reaction between carbon and hydrogen. The reaction products were sensitive to both pressure and temperature, as shown in Figure 6.13. In another study, a thermally produced beam of atomic hydrogen was reacted on a carbon target at temperatures between 30 and 950°C [40]. The reaction products isolated on a liquid helium finger and analyzed by gas chromatography were found to be mostly CH_4 with a small

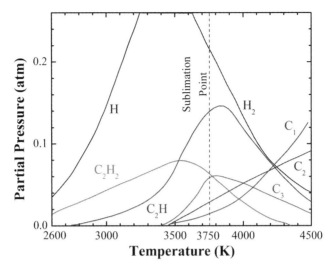

FIGURE 6.13   Equilibrium diagram of the C + H$_2$ system at 1 atm pressure. *Source*: Adapted with permission from Baddour and Iwasyk [39]. (See color insert.)

amount of C$_2$H$_6$ and C$_3$H$_8$. The content of formed hydrocarbon increased with temperature.

A more recent study examined the interaction between carbon and hydrogen atoms on a Ru(0001) surface using scanning tunneling microscopy (STM), density functional theory (DFT), and STM image calculations [41]. Formation of CH species by reaction between the adsorbed H and C was observed to occur readily at 100 K. When the coverage of H was increased, new complexes of the form of CH + nH (n = 1, 2, 3) were observed. These complexes were suggested as possible precursors for further hydrogenation reactions. DFT analysis indicated a considerable energy barrier for the CH + H → CH$_2$ reaction.

6.4.2 Reaction between Solid Carbon and Hydrogen

The reaction between carbon atom and hydrogen atom or molecule serves as a first step toward understanding reaction between solid carbon materials such as graphite, graphene, carbon nanotubes, and fullerenes. The interaction between hydrogen and solid carbon materials can vary from physical (weak) to chemical (strong) in nature. This topic will be addressed partly in Chapter 7, which focuses on various physical storage materials including carbon-based materials.

In this section, we will address briefly chemical reactions between solid carbon materials and hydrogen. While the detailed structures of the carbon

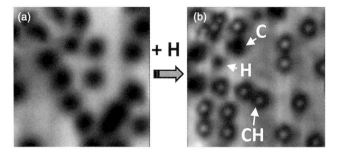

FIGURE 6.14 STM images (4×4 nm^2) of Ru(0001) acquired at $T = 6$ K. (a) Surface containing approximately 0.03 mL of C prepared by segregation from the bulk. The C atoms appear as depressions (black spots). (b) After introducing H atoms (from water dissociation in this experiment), C is converted to CH (bright protrusion surrounded by a dark ring). A similar transformation occurs with H obtained from H$_2$ dissociation. Individual nonreacted H atoms appear as smaller dark spots. Tunneling condition in (a) is $V_{sample} = 50$ mV and $I_t = 295$ pA, and (b) $V_{sample} = 9$ mV and $I_t = 495$ pA. The total z scale is adjusted to be 50 pm in both images. *Source*: Reproduced with permission from Shimizu et al. [41].

materials can vary substantially, the fundamental chemistry for their reaction with hydrogen is similar. As an example, the reaction of single-walled carbon nanotubes (SWNTs) with hydrogen gas studied in the temperature range of 400–550°C and at hydrogen pressure of 50 bar showed that hydrogenation of nanotubes was observed for samples treated at 400–450°C with about 1/3 of carbon atoms forming covalent C–H bonds, whereas hydrogen treatment at higher temperatures (550°C) occurred as an etching process, which was associated with the formation of light hydrocarbons [42]. The etching reaction of hydrogen on the edges of nanotubes was preferable compared with direct etching of the nanotube wall and starts at lower temperatures (400–450°C). Small hydrocarbon molecules such as CH$_4$ could be formed, especially at higher temperatures like 550°C. The reactions were likely facilitated by Fe nanoparticles that acted as catalysts for hydrogen dissociation. While this example demonstrates that it is possible to produce hydrocarbons from reactions of solid carbon such as CNTs with hydrogen, it is unclear yet if such reactions can become practically useful in storing hydrogen. Further studies are needed to understand the yield, mechanism, kinetics, and energy balance.

6.4.3 Reaction between Carbon Dioxide and Hydrogen

CO$_2$ is usually the major by-product of hydrogen release or generation from hydrocarbons, as discussed in Chapter 2. In order to recycle or reuse the carbon, it would be ideal to hydrogenate the CO$_2$ back to hydrocarbons. This is a demanding task, similar to hydrogenation of carbon directly. However, in principle, this can be accomplished, usually assisted with catalysis. Products of reaction between CO_2 and hydrogen include hydrocarbons such as methane and alcohols, for example methanol. An excellent review on this topic has appeared recently [43]. The basic steps for catalytic hydrocarbon generation involve conversion of CO_2 into CO followed by dissociation of CO into C and O and finally step-wise attachment of H to C. However, the detailed reaction mechanisms and intermediates are still not well understood, despite extensive studies.

For example, hydrogenation of CO_2 into CO is an important first step towards CO_2 reduction and hydrogen storage. This can be accomplished via the reserve water gas shift (RWGS) reaction:

$$CO_2 + H_2 = CO + H_2O, \Delta H_{298K} = 41.2 \text{ kJ} \cdot \text{mol}^{-1}. \tag{6.15}$$

This reaction is effective mainly with assistance of catalysts, such as Ni/CeO_2, and supported Cu, Fe, Pt, Ru, and Rh. The reaction mechanism has been examined extensively for Cu-based catalysts, which in itself is still controversial [43]. Two major mechanisms have been proposed: redox and formate decomposition. The redox mechanism is modeled by the following reactions:

$$CO_2 + 2Cu^0 \rightarrow Cu_2O \tag{6.16}$$

$$H_2 + Cu_2O \rightarrow 2Cu^0 + H_2O. \tag{6.17}$$

Cu^0 atoms are active in CO_2 dissociation and the reduction of the oxidized Cu catalyst has to be faster than the oxidation process. Hydrogen is proposed to be the reducing reagent without direct participation in the formation of intermediates [44]. The second model based on formate decomposition suggests that CO is formed from decomposition of formate intermediates derived from the association of hydrogen with CO_2 [45,46]. Different mechanisms have been proposed for the reaction with Pd- or Pt-based catalysts. For example, in a study involving Pd/Al_2O_3 catalysts and supercritical mixture of CO_2 and H_2, the formation of surface species such as carbonate, formate, and CO has been indicated based on infrared spectra [47].

Another important relevant reaction is methanation of CO_2, which is thermodynamically favorable with $\Delta G_{298\,K} = -130.8$ kJ·mol^{-1}:

$$CO_2 + 4H_2 \rightarrow CH_4 + 2\,H_2O, \Delta H_{298\,K} = 252.9 \text{ kJ} \cdot \text{mol}^{-1}. \tag{6.18}$$

This reaction is ideal since it produces CH_4, which can be easily used with current infrastructure and the by-product is water. The reaction is kinetically limited because it requires eight electrons to fully oxide carbon in CO_2 to methane, which requires catalysts to achieve acceptable rates and selectivities [48]. Common catalysts for this reaction include Ni, Ru, and Rh supported on various oxides such as SiO_2, TiO_2, Al_2O_3, and CeO_2. Figure 6.15 shows representative TEM and EDS images of Pd–Mg/SiO_2 catalysts [48]. The results show that the as-synthesized Pd–Mg/SiO_2 catalyst after calcination at 550°C in air for 6 hours contains well-dispersed electron-dense particles identified as containing Pd by EDS within a matrix of less dense, noncrystalline silica shells, with a certain degree of aggregation of Pd particles. After reaction with CO_2 and hydrogen for 10 hours, the particles remain well dispersed even though more larger Pd particles appeared, indicating sintering of some particles that are probably not fully encapsulated by the Mg/Si oxide.

The reaction mechanism is still not well established and the proposed mechanisms fall into two general categories. The first one involves the conversion of CO_2 to CO before methanation [49–51], while the second one involves direct hydrogenation of CO_2 to methane without forming CO as an intermediate [52, 53]. Even for mathanation of CO, there is no consensus on the kinetics and mechanism. It has been suggested that the rate-limiting step is either the formation of a CH_xO intermediate and its hydrogenation or the formation of surface carbon via CO dissociation and its interaction with hydrogen [51, 54].

To gain insight into the reaction mechanism, it is useful to determine reaction intermediates. For instance, kinetic studies based on steady-state transient measurements conducted on Ru/TiO_2 catalyst have identified several reaction intermediates [50]. Hydrogenation of CO as a key intermediate leads to methane formation. Formates, as intermediates for CO formation, are bound strongly on the support in equilibrium and become active species at the metal and support interface. A proposed mechanism for the formation of the formate thorough a carbonate species is shown in Figure 6.16.

Similarly, other hydrocarbons can be produced based on reaction between CO_2 and hydrogen. The reactions are often divided into two categories: methanol mediated and nonmethanol mediated. In the methanol-mediated approach, CO_2 and H_2 react over Cu–Zn-based catalysts to produce methanol, which is subsequently converted into other hydrocarbons such as gasoline. Light alkanes are usually generated as major products due to further catalytic hydrogenation of the alkenes. In the nonmethanol-mediated approach, CO_2 hydrogenation proceeds in two steps: RWGS reaction, which converts CO_2 into CO, and then the Fisher–Tropsch (FT) reaction, which converts CO into hydrocarbons via further hydrogenation.

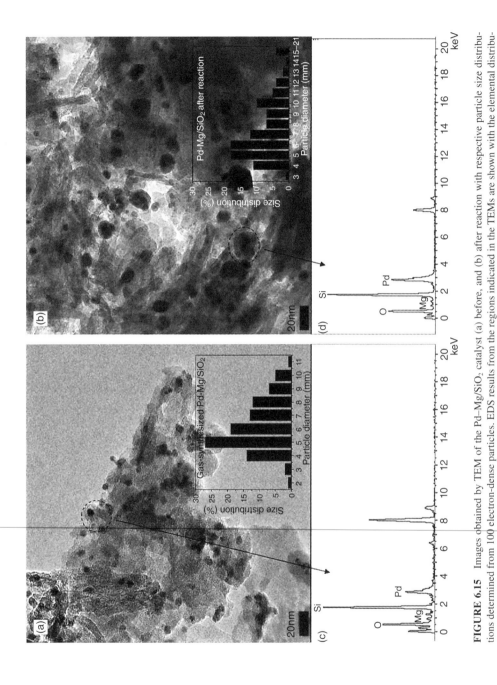

FIGURE 6.15 Images obtained by TEM of the Pd–Mg/SiO$_2$ catalyst (a) before, and (b) after reaction with respective particle size distributions determined from 100 electron-dense particles. EDS results from the regions indicated in the TEMs are shown with the elemental distributions. (c) and (d) Reproduced with permission from Park and McFarland [48].

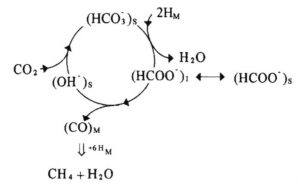

FIGURE 6.16 Schematic illustration of the CO_2 methanation process via hydrogenation. S stands for the support, M for the metal, and I for the metal-support interface. *Source*: Adapted from Marwood et al. [50].

For direct methanol generation from CO_2 hydrogenation, the reaction is given as:

$$CO_2 + 3\,H_2 \rightarrow CH_3OH + H_2O \; \Delta H_{298K} = -49.5 \; kJ \cdot mol^{-1}. \quad (6.19)$$

Based on thermodynamics, a decrease in temperature and an increase in pressure would favor methanol formation. By-products for methanol synthesis include CO, hydrocarbons, and higher alcohols. Thus, a highly selective catalyst is needed to avoid the formation of undesired by-products for methanol synthesis. Among the many metal-based catalysts, Cu remains the primary active catalyst component, together with various modifiers, such as Zn, Zr, Al, Ce, Si, Ti, B, and Cr. The catalysts are usually dispersed on oxide supports, such as ZnO and ZrO_2, which can play an important role in the reaction by affecting the formation and stability of the catalysts, as well as interaction between catalysts and promoters. Methanol selectivity is strongly dependent on the specific catalysts and supports used, and in some cases, near 100% selectivity is reached [55]. Despite extensive studies, the reaction mechanism of methanol synthesis is still not well understood. One model suggests that the reaction is occurring at the interfaces of Cu and oxides with CO_2 adsorbed on the oxides and H_2 dissociating on Cu [56].

6.5 SUMMARY

Metal hydrides represent one of the most promising materials for hydrogen storage. They usually have very high volumetric density but low gravimetric density. The advantages of metal hydrides include low pressure hydrogen adsorption and storage, as well as less rigorous requirement for containers. However, the thermodynamics and kinetics of metal hydrides present the greatest challenges since the formation of metal hydride and dehydrogenation process involve the formation or breaking of metal–hydrogen bonds. In most cases, high temperatures around 120–300°C are required to release the hydrogen, and the hydrogen adsorption and desorption rates are very slow. Sometimes, metal hydrides with high hydrogen content cannot adsorb and desorb hydrogen reversibly. Therefore, searching for new lightweight metal hydrides and tuning their thermodynamic and kinetic properties are the main research directions in this field.

As on the possibility of using hydrocarbons as a means for hydrogen storage, it is clear that while the potential exists, it is uncertain how effective this can be on a large scale. Further research is needed to better understand the reaction mechanisms as well as engineering issues for practical applications. This approach still involves CO_2, which is an environmental concern, but is attractive since hydrocarbons can be handled easily using existing infrastructures.

REFERENCES

1. Züttel, A., Remhof, A., Borgschulte, A., Friedrichs, O. Hydrogen: The future energy carrier. *Philosophical Transactions of the Royal Society A* **2010**, *368*, 3329–3342.

2. Schlapbach, L., Anderson, I., Burger, J.P. Hydrogen in metals. In K.H.J Buschow, ed., *Electronic and Magnetic Properties of Metals and Ceramics Part II*, VCH, Weinheim, 1994, Vol. 3B, p .271.

3. Zuttel, A. Materials for hydrogen srorage. *Materials Today* **2003**, *6*, 24–33.

4. Schlapbach, L., Zuttel, A. Hydrogen-storage materials for mobile applications. *Nature* **2001**, *414*(6861), 353–358.

5. Mintz, M.H., Zeiri, Y. Hydriding kinetics of powders. *Journal of Alloys and Compounds* **1994**, *216*(2), 159–175.

6. Avrami, M. Kinetics of phase change. I. General theory. *Journal of Chemical Physics* **1939**, *7*, 1103–1112.

7. http://www1.eere.energy.gov/hydrogenandfuelcells/storage/pdfs/targets_onboard _hydro_storage_explanation.pdf (last accessed December 19, 2013).

8. Broom, D.P. *Hydrogen Storage Materials: The Characterisation of Their Storage Properties*, Springer, London, 2011.

9. Hirscher, M. *Handbook of Hydrogen Storage: New Materials for Future Energy Storage*, WILEY-VCH, Weinheim, 2010.

10. Züttel, A., Borgschulte, A., Schlapbach, L. *Hydrogen as a Future Energy Carrier*, Wiley-VCH, Weinheim, 2008.

11. Stampfer, J.F.J., Holley, C.E.J., Suttle, J.F. The magnesium-hydrogen system. *Journal of the American Chemical Society* **1960**, *82*, 3504–350.

12. Dornheim, M., Eigen, N., Barkhordarian, G., Klassen, T., Bormann, R. Tailoring hydrogen storage materials towards application. *Advanced Engineering Materials* **2006**, *8*(5), 377–385.

13. Ross, D.K. Hydrogen storage: The major technological barrier to the development. of hydrogen fuel cell cars. *Vacuum* **2006**, *80*(10), 1084–1089.

14. Huot, J., Liang, G., Schulz, R., Mechanically alloyed metal hydride systems. *Applied Physics A* **2001**, *72*, 187–195.

15. Barkhordarian, G., Klassen, T., Bormann, R., Kinetic investigation of the effect of milling time on the hydrogen sorption reaction of magnesium catalyzed with different Nb_2O_5 contents. *Journal of Alloys and Compounds* **2006**, *407*, 249–255.

16. Sorensen, B. *Hydrogen and Fuel Cells*, Elsevier Academic Press, Amsterdam, 2005.

17. Sandrock, G. A panoramic overview of hydrogen storage alloys from a gas reaction point of view. *Journal of Alloys and Compounds* **1999**, *293–295*, 877–888.

18. Sorensen, B., *Hydrogen and Fuel Cells*, 3rd ed., Elsevier Academic Press, Amsterdam, 2004.

19. Bogdanovic, B., Schwickardi, M. Ti-doped alkali metal aluminium hydrides as potential novel reversible hydrogen storage materials. *Journal of Alloys and Compounds* **1997**, *253–254*, 1–9.

20. Chen, P., Xiong, Z., Luo, J.Z., Lin, J., Tan, K.L. Interaction of hydrogen with metal nitrides and imides. *Nature* **2002**, *420*, 302–304.

21. Jensen, C.M., Gross, K.J. Development of catalytically enhanced sodium aluminium hydride as a hydrogen-storage material. *Applied Physics A* **2001**, *72*, 213–219.

22. Berube, V., Radtke, G., Dresselhaus, M., Chen, G. Size effects on the hydrogen storage properties of nanostructured metal hydrides: A review. *International Journal of Energy Research* **2007**, *31*(6–7), 637–663.

23. Zaluski, L., Zaluska, A., Tessier, P., StromOlsen, J.O., Schulz, R. Nanocrystalline hydrogen absorbing alloys. In *Metastable, Mechanically Alloyed and Nanocrystalline Materials, Pts 1 and 2*, 1996, Vol. 225, pp. 853–858.

24. Zaluski, L., Zaluska, A., StromOlsen, J.O. Nanocrystalline metal hydrides. *Journal of Alloys and Compounds* **1997**, *253*, 70–79.

25. Zaluska, A., Zaluski, L., Strom-Olsen, J.O. Structure, catalysis and atomic reactions on the nano-scale: A systematic approach to metal hydrides for hydrogen storage. *Applied Physics A: Materials Science & Processing* **2001**, *72*(2), 157–165.

26. Sakintuna, B., Lamari-Darkrim, F., Hirscher, M. Metal hydride materials for solid hydrogen storage: A review. *International Journal of Hydrogen Energy* **2007**, *32*, 1121–1140.

27. Higuchi, K., Kajioka, H., Toiyama, K., Fujii, H., Orimo, S., Kikuchi, Y. In situ study of hydriding-dehydriding properties in some Pd/Mg thin films with different degree of Mg crystallization. *Journal of Alloys and Compounds* **1999**, *295*, 484–489.

28. Higuchi, K., Yamamoto, K., Kajioka, H., Toiyama, K., Honda, M., Orimo, S., Fujii, H. Remarkable hydrogen storage properties in three-layered Pd/Mg/Pd thin films. *Journal of Alloys and Compounds* **2002**, *330*, 526–530.

29. Akyildiz, H., Ozenbas, M., Ozturk, T. Hydrogen absorption in magnesium based crystalline thin films. *International Journal of Hydrogen Energy* **2006**, *31*(10), 1379–1383.

30. Wagemans, R.W.P., van Lenthe, J.H., de Jongh, P.E., van Dillen, A.J., de Jong, K.P. Hydrogen storage in magnesium clusters: Quantum chemical study. *Journal of the American Chemical Society* **2005**, *127*(47), 16675–16680.

31. He, Y.-P., Zhao, Y.-P. Hydrogen storage and cycling properties of Vanadium decorated Mg nanoblade array on Ti coated Si substrate. *Nanotechnology* **2009**, *20*, 204008.

32. Checchetto, R., Bazzanella, N., Miotello, A., Mengucci, P. Deuterium storage in Mg-Nb films. *Journal of Alloys and Compounds* **2005**, *404*, 461–464.

33. Huot, J., Liang, G., Boily, S., Van Neste, A., Schulz, R. Structural study and hydrogen sorption kinetics of ball-milled magnesium hydride. *Journal of Alloys and Compounds* **1999**, *293*, 495–500.

34. Berube, V., Dresselhaus, M.S., Chen, G. Entropy stabilization of deformed regions characterized by an excess volume for hydrogen storage applications. *International Journal of Hydrogen Energy* **2009**, *34*(4), 1862–1872.

35. Berube, V., Chen, G., Dresselhaus, M.S. Impact of nanostructuring on the enthalpy of formation of metal hydrides. *International Journal of Hydrogen Energy* **2008**, *33*(15), 4122–4131.

36. Kim, K.C., Dai, B., Johnson, J.K., Sholl, D.S. Assessing nanoparticle size effects on metal hydride thermodynamics using the Wulff construction. *Nanotechnology* **2009**, *20*(20).

37. Horvath, J., Birringer, R., Gleiter, H. Diffusion in nanocrystalline material. *Solid State Communications* **1987**, *62*(5), 319–322.

38. Frobose, K., Kolbe, F., Jackle, J. Size dependence of self-diffusion in a dense hard-disc liquid. *Journal of Physics: Condensed Matter* **2000**, *12*(29), 6563–6573.

39. Baddour, R.F., Iwasyk, J.M. Reactions between elemental carbon and hydrogen at temperatures above 2800 degrees K. *Industrial & Engineering Chemistry Process Design and Development* **1962**, *1*(3), 169–&.

40. Snelson, A. Reaction of atomic-hydrogen with carbon. *Advances in Chemistry Series* **1974**, *131*, 54–71.

41. Shimizu, T.K., Mugarza, A., Cerda, J.I., Salmeron, M. Structure and reactions of carbon and hydrogen on Ru(0001): A scanning tunneling microscopy study. *Journal of Chemical Physics* **2008**, *129*(24), 2441031–2441037.

42. Talyzin, A.V., Luzan, S., Anoshkin, I.V., Nasibulin, A.G., Jiang, H., Kauppinen, E.I., Mikoushkin, V.M., Shnitov, V.V., Marchenko, D.E., Noreus, D. Hydrogenation, purification, and unzipping of carbon nanotubes by reaction with molecular hydrogen: Road to graphane nanoribbons. *ACS Nano* **2011**, *5*(6), 5132–5140.

43. Wang, W., Wang, S.P., Ma, X.B., Gong, J.L. Recent advances in catalytic hydrogenation of carbon dioxide. *Chemical Society Reviews* **2011**, *40*(7), 3703–3727.

44. Chen, C.S., Wu, J.H., Lai, T.W. Carbon dioxide hydrogenation on Cu nanoparticles. *Journal of Physical Chemistry C* **2010**, *114*(35), 15021–15028.

45. Boccuzzi, F., Chiorino, A., Martra, G., Gargano, M., Ravasio, N., Carrozzini, B. Preparation, characterization, and activity of Cu/TiO$_2$ catalysts .1. Influence of the preparation

method on the dispersion of copper in Cu/TiO_2. *Journal of Catalysis* **1997**, *165*(2), 129–139.

46. Coloma, F., Marquez, F., Rochester, C.H., Anderson, J.A. Determination of the nature and reactivity of copper sites in $Cu-TiO_2$ catalysts. *Physical Chemistry Chemical Physics* **2000**, *2*(22), 5320–5327.

47. Arunajatesan, V., Subramaniam, B., Hutchenson, K.W., Herkes, F.E. In situ FTIR investigations of reverse water gas shift reaction activity at supercritical conditions. *Chemical Engineering Science* **2007**, *62*(18–20), 5062–5069.

48. Park, J.N., McFarland, E.W. A highly dispersed $Pd-Mg/SiO_2$ catalyst active for methanation of CO2. *Journal of Catalysis* **2009**, *266*(1), 92–97.

49. Peebles, D.E., Goodman, D.W., White, J.M. Methanation of carbon-dioxide on Ni(100) and the effects of surface modifiers. *Journal of Physical Chemistry* **1983**, *87*(22), 4378–4387.

50. Marwood, M., Doepper, R., Renken, A., In-situ surface and gas phase analysis for kinetic studies under transient conditions: The catalytic hydrogenation of CO_2. *Applied Catalysis A: General* **1997**, *151*(1), 223–246.

51. Lapidus, A.L., Gaidai, N.A., Nekrasov, N.V., Tishkova, L.A., Agafonov, Y.A., Myshenkova, T.N. The mechanism of carbon dioxide hydrogenation on copper and nickel catalysts. *Petroleum Chemistry* **2007**, *47*(2), 75–82.

52. Fujita, S., Terunuma, H., Kobayashi, H., Takezawa, N. Methanation of carbon-monoxide and carbon-dioxide over nickel-catalyst under the traisnet state. *Reaction Kinetics and Catalysis Letters* **1987**, *33*(1), 179–184.

53. Schild, C., Wokaun, A., Baiker, A. On the mechanim of CO and CO_2 hydrogenation reactions on zirconia-supported catalysts: A diffuse reflectance FTIR study. 2. Surface species on copper zirconia catalysts: Implications for methanol synthesis selectivity. *Journal of Molecular Catalysis* **1990**, *63*(2), 243–254.

54. Sehested, J., Dahl, S., Jacobsen, J., Rostrup-Nielsen, J.R., Methanation of CO over nickel: Mechanism and kinetics at high H_2/CO ratios. *Journal of Physical Chemistry B* **2005**, *109*(6), 2432–2438.

55. Sloczynski, J., Grabowski, R., Kozlowska, A., Olszewski, P., Stoch, J., Skrzypek, J., Lachowska, M. Catalytic activity of the $M/(3ZnO \cdot ZrO_2)$ system (M = Cu, Ag, Au) in the hydrogenation of CO_2 to methanol. *Applied Catalysis a-General* **2004**, *278*(1), 11–23.

56. Fisher, I.A., Bell, A.T. In-situ infrared study of methanol synthesis from H_2/CO_2 over Cu/SiO_2 and $Cu/ZrO_2/SiO_2$. *Journal of Catalysis* **1997**, *172*(1), 222–237.

7

Physical Storage Using Nanostructured and Porous Materials

7.1 PHYSICAL STORAGE USING NANOSTRUCTURES

Besides compression, liquification, and chemical storage, hydrogen can also be stored based on physical absorption using various carbon-based or noncarbon-based materials. Nanostructured materials are particularly attractive due to their large surface-to-volume ratio. Nanomaterials used for hydrogen storage based on physisorption include carbon nanostructures, organic polymers, inorganic nanostructures, and composite structures. For ease of discussion, we will divide them into two categories: carbon based and noncarbon based, and discuss each separately next.

Figure 7.1 shows schematic summary of some of the different physical storage methods as compared with chemical storage. For storage based on physisorption, large surface area is critical. Nanostructured materials are therefore ideally suited for such purpose.

7.1.1 Carbon Nanostructures

Carbon nanostructures, particularly carbon nanotubes (CNTs), fullerenes, nanofibers, and, more recently, graphenes, have been studied for hydrogen storage [1, 2]. They are attractive due to a combination of good adsorption ability, high specific surface area, porous microstructure, and low mass

Hydrogen Generation, Storage, and Utilization, First Edition. Jin Zhong Zhang, Jinghong Li, Yat Li, and Yiping Zhao.
© 2014 John Wiley & Sons, Inc. Published 2014 by John Wiley & Sons, Inc.

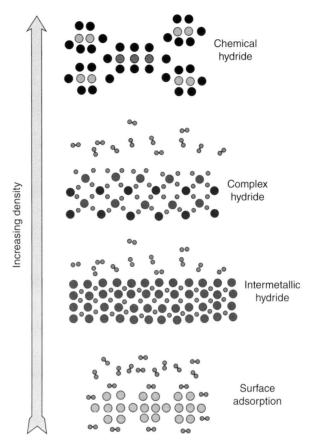

FIGURE 7.1 Hydrogen storage density in physisorbed materials, metal/complex, and chemical hydrides. *Source*: Reproduced with permission from Niemann et al. [1]. (See color insert.)

density. The actual mechanism of hydrogen storage is still not well understood, but is considered to involve either physisorption via van der Waals attractive forces or chemisorption via strong interaction or bonding between hydrogen and carbon atoms. Physisorption of hydrogen limits the hydrogen-to-carbon ratio to less than one hydrogen atom per two carbon atoms (i.e., 4.2 mass %), whereas chemisorption allows a higher hydrogen to carbon ratio, with the ratio of two hydrogen atoms per one carbon atom reported in the case of polyethylene [3, 4]. Physisorbed hydrogen has a binding energy normally on the order of 0.1 eV, while chemisorbed hydrogen has C–H covalent bonding, with a binding energy of more than 2–3 eV.

The first report of hydrogen storage using carbon nanotubes appeared in 1997 [5], which stimulated significant interest in this topic. Both single-walled and multiwalled CNTs have been studied, with hydrogen storage values between 0.25 and 56 wt% reported under various experimental conditions [2]. The very high values reported (30–60 wt%) have been questioned and were later considered as possibly the result of measurement errors. However, storage capacity of a few wt% is still considered as potentially useful. Very recently, the hydrogen storage capacities have been reported at ambient temperature to be 1.7, 1.85, 3.0, and 2.0 wt% for double-wall CNTs (DWCNTs) loaded with 1, 1.0, 2.0, and 3.0 wt% Pd, respectively [6]. The hydrogen storage capacity can be enhanced by loading with Pd nanoparticles and selecting a suitable content, and the sorption was attributed to the chemical reaction between the atomic hydrogen and the dangling bonds of the DWCNTs. Figure 7.2 shows TEM bright-field micrographs of loaded and unloaded DWCNTs with different weight percentage of Pd and corresponding selected area electron diffraction patterns. In another study, Pt decoration, in conjunction with acidic etching, was found to substantially enhance hydrogen storage to a capacity of about 2.7 wt% [7]. It was suggested that the acidic etching increased the surface defect density and led to opening up the

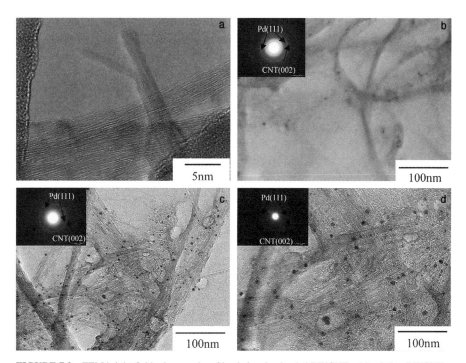

FIGURE 7.2   TEM bright-field micrographs of loaded and unloaded DWCNTs: (a) pristine DWCNTs; (b) 1 wt%Pd/DWCNTs; (c) 2 wt%Pd/DWCNTs; and (d) 3 wt%Pd/DWCNTs, with insets in (b), (c), and (d) showing the corresponding selected area electron diffraction patterns. *Source*: Reproduced with permission from Wu et al. [6].

caps of CNTs, resulting in an increase in the active adsorption site for physical sorption of H_2 while the Pt nanoparticles promoted chemical sorption of hydrogen via spillover mechanism that involves additives (e.g., Ru, Pt, and Pd) that act as catalysts for dissociation of hydrogen molecules to hydrogen atoms, which move from the catalytic sites to the surface of CNTs and finally become adsorbed.

Fullerenes are molecules composed entirely of carbon, in the form of hollow spheres, ellipsoids, or tubes. CNTs discussed earlier belong to the family of fullerenes with cylindrical shapes, also called buckytubes. A special spherical fullerene structure with 60 carbon atoms is called buckminsterfullerene or buckyball, as it resembles the ball used in football (soccer) [8]. Buckminsterfullerene or C_{60} has a cage-like fused-ring structure (truncated icosahedron), made of 20 hexagons and 12 pentagons, with a carbon atom at each vertex of each polygon and a bond along each polygon edge. It has been considered as a potential hydrogen storage material because of its ability to react with hydrogen via the C=C double bonds. Theoretically, $C_{60}H_{60}$ can be formed that would correspond to a hydrogen content of ~7.7 wt% [1]. Experimentally, this is challenging to realize due to the requirement of very high temperature, about 823–873 K [9]. Strategies have been developed to overcome the challenge, including doping of C_{60} using metal atoms such as Li [10–13]. One example is the recent work on Li-doped fullerene ($Li_x–C_{60}–H_y$) that is capable of reversibly storing hydrogen through chemisorption at elevated temperatures and pressures [13]. This system is unique in that hydrogen is closely associated with lithium and carbon upon rehydrogenation of the material and that the weight percent of H_2 stored in the material is closely linked to the stoichiometric ratio of Li:C_{60} in the material. Under optimized conditions, a Li-doped fullerene with a Li:C_{60} mole ratio of 6:1 can reversibly desorb up to 5 wt% H_2 with an onset temperature of ~270°C, which is significantly lower than the desorption temperature of hydrogenated fullerenes ($C_{60}H_x$) and pure lithium hydride (decomposition temperature 500–600 and 670°C, respectively). However, the $Li_x–C_{60}–H_y$ system still suffers from the same drawbacks as typical hydrogenated fullerenes (high desorption temperature and release of hydrocarbons) because the fullerene cage remains mostly intact and is only slightly modified during multiple hydrogen desorption/absorption cycles.

Graphene, a substance made of pure carbon with atoms arranged in a regular hexagonal pattern similar to graphite but in a one-atom thick sheet, has been considered for hydrogen storage as well, even though the studies have been limited so far. It potentially can have a very high hydrogen storage density. Graphene has been converted into graphane reversibly using a stream of hydrogen atoms, and graphane can release the stored hydrogen by heating

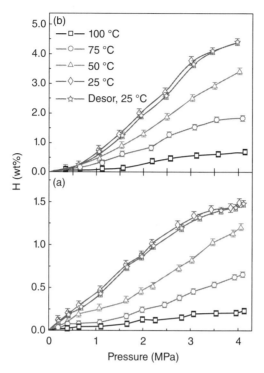

FIGURE 7.3   Pressure—composition isotherms of (a) N-doped hydrogen exfoliated graphene (N-HEG) and (b) Pd-decorated H-HEG (Pd–N–HEG) in the temperature range 25 – 100°C and 0.1 – 4 MPa pressure. *Source*: Reproduced with permission from Parambhath et al. [15].

at 450°C [14]. In particular, recent experimental studies have found that doping or composite structures of graphene or graphene oxide (GO) with metal or nonmetal elements or nanoparticles can substantially improve their hydrogen storage capacity [15–17]. For instance, a high hydrogen storage capacity for Pd decorated, N-doped hydrogen exfoliated graphene nanocomposite is demonstrated under moderate temperature and pressure [15]. Figure 7.3 shows pressure—composition isotherms of N-doped hydrogen exfoliated graphene (N-HEG) and Pd-decorated H-HEG (Pd–N–HEG) in the temperature range 25 – 100°C and 0.1 – 4 MPa pressure. An increase of 66% is achieved by N-doping in the hydrogen uptake capacity of hydrogen exfoliated graphene at room temperature and 2 MPa pressure, with a further enhancement by 124% attained in the hydrogen uptake capacity by Pd nanoparticle decoration over N-doped graphene. The high hydrogen uptake capacities obtained was attributed to interplay between the catalyst support and catalyst particles. Similarly, enhanced hydrogen storage capacity of pristine graphene nanoplatelets has been found by others and attributed to an increase in the hydrogen spillover effect and the binding energy between metal nanoparticles and supporting material facilitated by nitrogen doping.[18]

Several recent theoretical or computational studies have also found that chemical doping or nanocomposite can improve the hydrogen storage performance of graphene or GO [19–22]. For example, detailed first-principles calculations based on density functional theory have been carried out on graphene with Li atoms with the objective to determine how the Li coverage pattern affects the hydrogen storage capacity [22]. Results indicate that hydrogen storage capacity can be increased to 16 wt% by adjusting the coverage of Li atoms on graphene to the (root 3 × root 3) pattern at both sides. This study demonstrates the importance of the details of the surface coverage of the metal atoms as well as the potential of metal-modified graphene for hydrogen storage.

Carbon nanofbers (CNFs) are cylindric carbon-based nanostructures with graphene layers arranged as stacked cones, cups, or plates. Carbon nanotubes discussed earlier are nanofibers with graphene layers wrapped into perfect cylinders. Similar to other carbon nanostructures, CNFs have been studied for hydrogen storage with encouraging results. For example, CNFs synthesized by a catalytic pyrolysis method can store 4 wt% or higher of hydrogen, similar to CNTs [23]. Similarly, turbostratic CNFs with a rough surface, open pore walls, and a defect structure, produced by the thermal decomposition of alcohol in the presence of an iron catalyst and a sulfur promoter at 1100°C under a nitrogen atmosphere in a vertical chemical vapor deposition reactor, showed hydrogen storage capacities 1.5 and 5 wt% for the as-produced and exfoliated forms, respectively [24]. The defects on the surface and expandable graphitic structure are considered important to increasing the hydrogen uptake. In a more recent study that compares the hydrogen adsorption capacity of different types of carbon nanofibers (platelet, fishbone, and ribbon) and amorphous carbon measured as a function of pressure and temperature, more graphitic/ordered carbon materials have been found to adsorb less hydrogen than the more amorphous ones, and functionalization (oxygen surface group incorporation) and Ni-modification considerably improve the hydrogen adsorption capacity [25]. Figure 7.4 shows representative TEM images of parent carbon materials: (a) amorphous carbon, (b) ribbon CNFs, (c) platelet CNFs, and (d) fishbone CNFs. The functionalization helps the development of pores and accessibility of internal surface while Ni-modification enhances the spillover effect, which involves the initial H_2 adsorption and dissociation (metal catalyzed process) followed by the dissociated H migration through the metal and its anchorage to the carbonaceous structure [26].

Activated carbon (AC) is a highly porous, modified synthetic carbon that contains crystalized graphite and amorphous carbon with a high specific

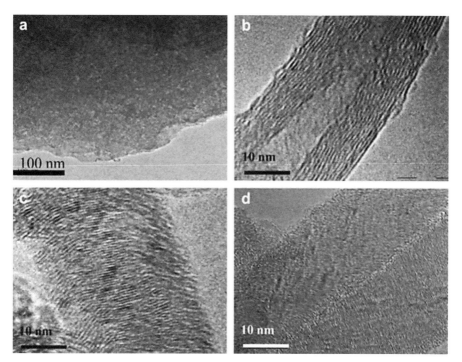

FIGURE 7.4   Representative TEM images of parent carbon materials: (a) amorphous carbon, (b) ribbon CNFs, (c) platelet CNFs and (d) fishbone CNFs. *Source*: Reproduced with permission from Jimenez et al. [25].

surface area. It is potentially useful for hydrogen storage due to its low cost and accessibility on a commercial scale [27]. For hydrogen storage, the rate and capacity of the activated carbon are influenced by its morphology and shape, that is, powder, fiber, and granular. For conventional AC, the hydrogen update is proportional to its surface area and pore volume, and high adsorption capacity is only achieved at very low cryogenic temperature and high pressure. Various types of commercial and modified AC have been extensively studied. For example, the capacity of hydrogen storage has been studied for electrospun-activated carbon fibers prepared by electrospinning and chemical activation based on the comparison with other carbon materials such as active carbon, single-walled carbon nanotube, and graphite [28]. The hydrogen adsorption capacity of chemically activated electrospun carbon fiber is better than that of other porous carbon materials, which is attributed to the optimized pore structure of electrospun-activated carbon fibers that might provide better sites for hydrogen adsorption than other carbon materials. Figure 7.5 shows some representative SEM images of several electrospun-activated carbon fibers, which all exhibit highly uniform diameters.

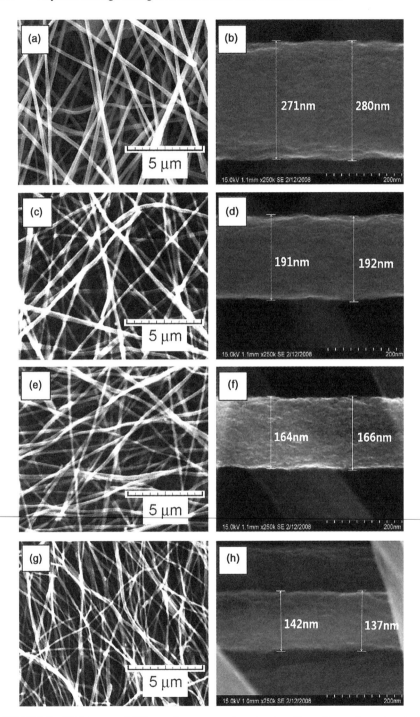

FIGURE 7.5 SEM images of different electrospun-activated carbon fibers. *Source*: Reproduced with permission from Im et al. [28].

In another study, ultrahigh surface area carbons (3000–3500 $m^2 \cdot g^{-1}$) have been obtained via chemical activation of polypyrrole with KOH and the carbon materials exhibit large pore volumes (up to similar to 2.6 $cm^3 \cdot g^{-1}$) and possess two pore systems: one of pores in the micropore range (similar to 1.2 nm) and the other in the small mesopore range (2.2–3.4 nm) [29]. Tuning of the carbon textural properties through the control of the activation parameters (temperature and amount of KOH) led to the generation of activated carbon that exhibits excellent hydrogen storage capacity of up to 7.03 wt% at −196°C and 20 bar, which is the highest ever reported for one-step ACs and among the best for any porous material. The gravimetric hydrogen uptake of the carbons translates to a very attractive volumetric density of up to 37 g $H_2 \cdot L^{-1}$ at 20 bar. These carbons exhibit excellent gravimetric and volumetric capacity due to the fact that their high porosity is not at the detriment of packing density.

In a very recent study, the effect of Pd nanoparticle doping of AC on hydrogen storage capacity has been studied [30]. Three ACs with an apparent surface area ranging from 2450 to 3200 m^2/g were doped with Pd nanoparticles at different levels within the range 1.3–10.0 wt%, and their excess hydrogen storage capacities were measured at 77 and 298 K at pressures up to 8 MPa. The hydrogen storage properties depend linearly on Pd content when hydrogen storage is carried out at 298 K and at pressures up to 1 MPa. At higher pressures, hydrogen storage depends on microporous volume so Pd addition does not bring capacity enhancement. Hydrogen storage at 77 K fundamentally depends on the specific micropore volume and consequently, Pd doping decreases hydrogen storage capacities by decreasing the specific micropore volume available for physisorption.

Other carbon-related but not pure carbon-based nanomaterals for hydrogen storage include pristine and modified polymers or composite structures involving organic polymers and metal-doped polymers [17, 31–33]. Some of these materials show improved hydrogen storage capacity compared with pristine carbon nanostructures.

7.1.2   Other Nanostructures and Microstructures

Some noncarbon-based nanostructures and microstructures have also been investigated for hydrogen storage, for example, glass capillary arrays [34–36], hollow glass microspheres [37, 38], and zeolites [39–43]. Glass capillary arrays have been developed as systems for safe infusion, storage, and controlled release of hydrogen gas, with storage pressures up to 1200 bar [36]. This technology enables the storage of a significantly higher amount of hydrogen than other approaches, and the main determinant in this storage technology is the pressure resistance of glass capillaries.

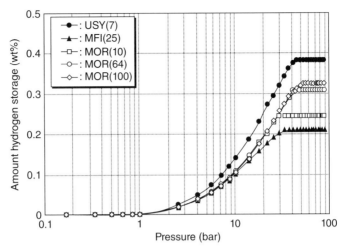

FIGURE 7.6   Hydrogen adsorption isotherms of various zeolites at 30°C. *Source*: Reproduced with permission from Chung [42].

In another study, the roles of the framework structure, surface area, and pore volume of microporous zeolites on hydrogen adsorption have been investigated using a high pressure dose of hydrogen at 30°C [42]. Figure 7.6 shows representative hydrogen adsorption isotherms on different microporous zeolites, which reached equilibrium after being dosed with 50 bar of hydrogen. The largest hydrogen adsorption was approximately 0.4 wt% on USY(7) zeolite. Although this storage capacity is insufficient to the target of DOE, it can be considered as a storage material of hydrogen with its modification by ion exchange and enlargement of pore volume, because the zeolites have a large pore volume and suitable channel diameter close to kinetic diameter of the hydrogen molecule (2.89 Å). The amount of hydrogen adsorption on mordenite (MOR) zeolites increased with increasing Si/Al molar ratio, which was achieved by dealumination. The amount of hydrogen adsorption increased linearly with increasing pore volume of the zeolites. The hydrogen adsorption behavior was found to be dependent mainly on the pore volume of the zeolites.

7.2   PHYSICAL STORAGE USING METAL-ORGANIC FRAMEWORKS

Metal–organic frameworks (MOFs) are a unique class of synthetic porous materials that have been demonstrated to store hydrogen. Due to the special characters of MOFs, we will discuss them in a separate section here.

MOFs are highly crystalline inorganic–organic hybrid structures that contain metal clusters or ions (secondary building units, or SBUs) as nodes and organic ligands as linkers. When guest molecules (solvent) occupying the pores are removed during solvent exchange and heating under vacuum, porous structure of MOFs can be achieved without destabilizing the frame, and hydrogen molecules can be adsorbed onto the surface of the pores by physisorption. Compared with traditional zeolites and porous carbon materials, MOFs have very high number of pores and large surface area that allow higher hydrogen uptake in a given volume. Thus, research interests in hydrogen storage using MOFs have been growing since 2003 when the first MOF-based hydrogen storage was introduced [44]. Since there are infinite geometric and chemical variations of MOFs based on different combinations of SBUs and linkers, many researches have explored what combination will provide the maximum hydrogen uptake by varying materials of metal ions and linkers.

In 2006, hydrogen storage concentrations of up to 7.5 wt% in MOF-74 has been achieved at a low temperature of 77 K [45]. In 2009, a higher storage concentration (10 wt%) at 77 bar (1117 psi) and 77 K with MOF NOTT-112 was reported [46]. Most studies of hydrogen storage in MOFs have been conducted at a temperature of 77 K and a pressure of 1 bar because such condition is commonly available and the binding energy between hydrogen and MOF is large compared with the thermal energy that can allow high hydrogen uptake capacity. The amount of hydrogen uptake depends on a number of factors, such as surface area, pore size, catenation, ligand structure, spillover, and sample purity.

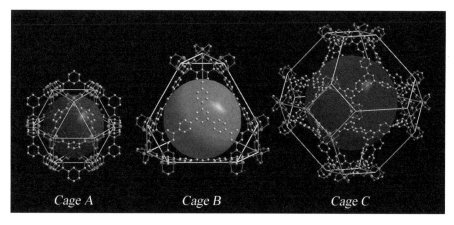

FIGURE 7.7   Different cages in the crystal structure of NOTT-112. Copper: blue–green; carbon: grey; oxygen: red. Water molecules and H atoms are omitted for clarity. *Source*: Reproduced with permission from Yan et al. [46]. (See color insert.)

For example, a recent study has examined the effects of structural modifications on the evolution of the crystal structure, pore characteristics, and H_2 capacities of MOF-5s and found that structural modifications can significantly influence the pore characteristics, and the specific surface areas of the MOF-5s decreased with the evolution of an ultrafine porosity [47]. These changes were correlated with an increase in the H_2 storage capacity of the MOF-5 (from 1.2 to 2.0 wt% at 196°C and 1 bar). The structural modifications also enhanced the thermal stability of the MOF-5s. In another study, porous carbon with hierarchical pore structure derived from highly crystalline MOFs (denoted as MOF-derived carbon or MDC) without any carbon source was found to display hierarchical pore structures with high ultramicroporosity, high specific surface area, and very high total pore volume, and thereby to exhibit reversible H_2 storage capacities at certain conditions that were better than those of previously reported porous carbons and MOFs, with a 3.25 wt% uptake for MDC-1 at 77 K and 1 bar [48]. This exceeded that of the benchmark materials and surpassed the performance of all other materials characterized to date. Figure 7.8 shows a comparison of the hydrogen storage capacity of the MDCs and isoreticular MOFS (IRMOFs) with some benchmark materials such as PCN-12 and ZTC.

7.3 CLATHRATE HYDRATES

Clathrate hydrates are a class of solid inclusion compounds in which guest molecules occupy cages formed from hydrogen-bonded water molecule networks. The usually unstable empty cages can be stabilized by inclusion of appropriately sized molecules. Clathrate hydrates of hydrogen often possess two different-sized cages to meet the necessary storage requirements. However, the higher pressures required, around 2 kbar, to produce the material makes it impractical. The synthesis pressure can be decreased by filling the larger cavity with tetrahydrofuran (THF) to stabilize the material, with the compromise of the potential storage capacity of the material [49]. In a related study, reversible hydrogen storage capacities in THF-containing binary-clathrate hydrates have been increased to $\sim$4 wt% at 270 K and modest pressures (12 MPa) by tuning their composition to allow the hydrogen guests to enter both the larger and the smaller cages, while retaining low pressure stability. The tuning mechanism is quite general and convenient, using water-soluble hydrate promoters and various small gaseous guests [50]. Figure 7.9 shows the hydrogen content as a function of THF concentration and a schematic diagram of hydrogen distribution in the cages of THF+H_2 hydrates. A subsequent study found that the use of inexpensive hydrogels as

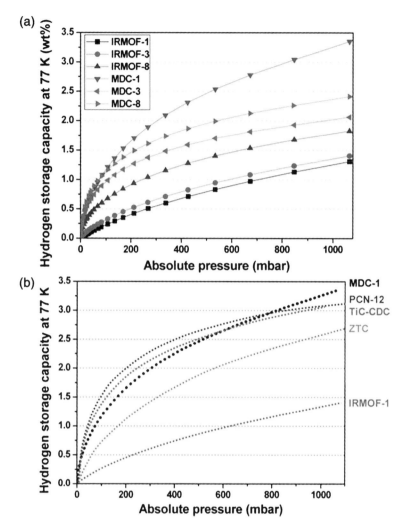

FIGURE 7.8   Hydrogen storage capacity at 77 K, and 1 bar of (a) the products and (b) the benchmark materials are shown for comparison. IRMOF stands for isoreticular MOF and MDC for MOD derived. *Source*: Reproduced with permission from Yang et al. [48]. (See color insert.)

supports can significantly improve H_2 enclathration kinetics and capacities in THF-H_2O clathrate hydrate with respect to bulk solutions, suggesting potential for accelerated gas-storage kinetics in clathrate-based technologies using polymer hydrogels [51].

Besides experimental studies, theoretical or computation studies have been carried to gain a better understanding of the hydrogen storage capacity and mechanism of calthrate hydrates [52, 53]. For instance, a very recent

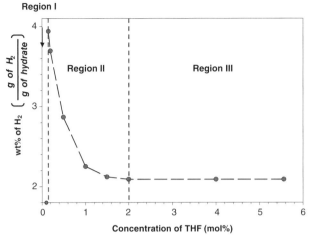

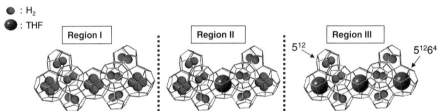

FIGURE 7.9 H_2 gas content as a function of THF concentration, and a schematic diagram of H_2 distribution in the cages of THF+H_2 hydrate. (H_2 gas content is calculated from g of H_2 per g of hydrate, and expressed as wt%.) In region III, H_2 molecules are only stored in small cages, while in region II, both small and large cages can store H_2 molecules. At the highly dilute THF concentrations of region I, H_2 molecules can still be stored in both cages, but extreme pressures (~2 kbar) are required to form the hydrates. Pure H_2 clathrate $(2H_2)_2 \cdot (4H_2) \cdot 17H_2O$ would have a 5.002 wt% H2 content. *Source*: Reproduced with permission from Lee et al. [50]. (See color insert.)

computational study based on first principles electronic structure calculations of the pentagonal dodecahedron, $(H_2O)_{20}$, (D-cage) and tetrakaidecahedron, $(H_2O)_{24}$, (T-cage) building blocks of structure I (sI) hydrate lattice suggest that these can accommodate up to a maximum of 5 and 7 guest hydrogen molecules, respectively [54]. For the pure hydrogen hydrate, Born–Oppenheimer molecular dynamics (BOMD) simulations of periodic (sI) hydrate lattices indicate that the guest molecules are released into the vapor phase via the hexagonal faces of the larger T-cages. The presence of methane in the larger T-cages was found to block this release, therefore suggesting possible means for stabilizing these coated clathrate hydrates and the potential enhancement of their hydrogen storage capacity.

7.4 SUMMARY

It is clear from the discussion earlier that each material considered for hydrogen storage has some strengths and limitations and none can meet all the desired criteria and target requirements yet. Compared with hydrogen generation and utilization, hydrogen storage faces more and greater challenges. Further research and development are needed, both in the experimental and theoretical fronts, to meet these challenges. Some fundamental issues still need to be addressed systematically at the atomic level in relation to both the thermodynamics and kinetics of hydrogen adsorption and desorption.

REFERENCES

1. Niemann, M.U., Srinivasan, S.S., Phani, A.R., Kumar, A., Goswami, D.Y., Stefanakos, E.K. Nanomaterials for hydrogen storage applications: A review. *Journal of Nanomaterials* **2008**, 1–9.

2. Lim, K.L., Kazemian, H., Yaakob, Z., Daud, W.R.W. Solid-state materials and methods for hydrogen storage: A critical review. *Chemical Engineering & Technology* **2010**, *33*(2), 213–226.

3. Sudan, P., Zuttel, A., Mauron, P., Emmenegger, C., Wenger, P., Schlapbach, L. Physisorption of hydrogen in single-walled carbon nanotubes. *Carbon* **2003**, *41*(12), 2377–2383.

4. Nijkamp, M.G., Raaymakers, J., van Dillen, A.J., de Jong, K.P. Hydrogen storage using physisorption: Materials demands. *Applied Physics a-Materials Science & Processing* **2001**, *72*(5), 619–623.

5. Dillon, A.C., Jones, K.M., Bekkedahl, T.A., Kiang, C.H., Bethune, D.S., Heben, M.J. Storage of hydrogen in single-walled carbon nanotubes. *Nature* **1997**, *386*(6623), 377–379.

6. Wu, H.M., Wexler, D., Liu, H. Effects of different palladium content loading on the hydrogen storage capacity of double-walled carbon nanotubes. *International Journal of Hydrogen Energy* **2012**, *37*(7), 5686–5690.

7. Tsai, P.J., Yang, C.H., Hsu, W.C., Tsai, W.T., Chang, J.K. Enhancing hydrogen storage on carbon nanotubes via hybrid chemical etching and Pt decoration employing supercritical carbon dioxide fluid. *International Journal of Hydrogen Energy* **2012**, *37*(8), 6714–6720.

8. Kroto, H.W., Heath, J.R., Obrien, S.C., Curl, R.F., Smalley, R.E. C-60: Buckminsterfullerene. *Nature* **1985**, *318*(6042), 162–163.

9. Peera, A.A., Alemany, L.B., Billups, W.E. Hydrogen storage in hydrofullerides. *Applied Physics A: Materials Science & Processing* **2004**, *78*(7), 995–1000.

10. Sun, Q., Jena, P., Wang, Q., Marquez, M. First-principles study of hydrogen storage on Li12C60. *Journal of the American Chemical Society* **2006**, *128*(30), 9741–9745.

11. Yoshida, A., Okuyama, T., Terada, T., Naito, S. Reversible hydrogen storage/release phenomena on lithium fulleride (LinC60) and their mechanistic investigation by solid-state NMR spectroscopy. *Journal of Materials Chemistry* **2011**, *21*(26), 9480–9482.

12. Paolone, A., Vico, F., Teocoli, F., Sanna, S., Palumbo, O., Cantelli, R., Knight, D.A., Teprovich, J.A., Zidan, R. Relaxation processes and structural changes in Li- and Na-Doped fulleranes for hydrogen storage. *Journal of Physical Chemistry C* **2012**, *116*(31), 16365–16370.

13. Teprovich, J.A., Wellons, M.S., Lascola, R., Hwang, S.J., Ward, P.A., Compton, R.N., Zidan, R. Synthesis and characterization of a lithium-doped fullerane (Li-x-C-60-H-y) for reversible hydrogen storage. *Nano Letters* **2012**, *12*(2), 582–589.

14. Elias, D.C., Nair, R.R., Mohiuddin, T.M.G., Morozov, S.V., Blake, P., Halsall, M.P., Ferrari, A.C., Boukhvalov, D.W., Katsnelson, M.I., Geim, A.K., Novoselov, K.S. Control of Graphene's properties by reversible hydrogenation: Evidence for graphane. *Science* **2009**, *323*(5914), 610–613.

15. Parambhath, V.B., Nagar, R., Ramaprabhu, S. Effect of nitrogen doping on hydrogen storage capacity of palladium decorated graphene. *Langmuir* **2012**, *28*(20), 7826–7833.

16. Xi, P.X., Chen, F.J., Xie, G.Q., Ma, C., Liu, H.Y., Shao, C.W., Wang, J., Xu, Z.H., Xu, X.M., Zeng, Z.Z. Surfactant free RGO/Pd nanocomposites as highly active heterogeneous catalysts for the hydrolytic dehydrogenation of ammonia borane for chemical hydrogen storage. *Nanoscale* **2012**, *4*(18), 5597–5601.

17. Zhang, F., Hou, C.Y., Zhang, Q.H., Wang, H.Z., Li, Y.G. Graphene sheets/cobalt nanocomposites as low-cost/high-performance catalysts for hydrogen generation. *Materials Chemistry and Physics* **2012**, *135*(2–3), 826–831.

18. Vinayan, B.P., Sethupathi, K., Ramaprabhu, S. Hydrogen storage studies of palladium decorated nitrogen doped graphene nanoplatelets. *Journal of Nanoscience and Nanotechnology* **2012**, *12*(8), 6608–6614.

19. Wu, M.H., Gao, Y., Zhang, Z.Y., Zeng, X.C. Edge-decorated graphene nanoribbons by scandium as hydrogen storage media. *Nanoscale* **2012**, *4*(3), 915–920.

20. Wang, V., Mizuseki, H., He, H.P., Chen, G., Zhang, S.L., Kawazoe, Y. Calcium-decorated graphene for hydrogen storage: A van der Waals density functional study. *Computational Materials Science* **2012**, *55*, 180–185.

21. Zhao, G.F., Li, Y., Liu, C.S., Wang, Y.L., Sun, J.M., Gu, Y.Z., Wang, Y.X., Zeng, Z. Boron nitride substrate-induced reversible hydrogen storage in bilayer solid matrix via interlayer spacing. *International Journal of Hydrogen Energy* **2012**, *37*(12), 9677–9687.

22. Zhou, W.W., Zhou, J.J., Shen, J.Q., Ouyang, C.Y., Shi, S.Q. First-principles study of high-capacity hydrogen storage on graphene with Li atoms. *Journal of Physics and Chemistry of Solids* **2012**, *73*(2), 245–251.

23. Cheng, H.M., Liu, C., Fan, Y.Y., Li, F., Su, G., Cong, H.T., He, L.L., Liu, M. Synthesis and hydrogen storage of carbon nanofibers and single-walled carbon nanotubes. *Zeitschrift Fur Metallkunde* **2000**, *91*(4), 306–310.

24. Wu, H.C., Li, Y.Y., Sakoda, A. Synthesis and hydrogen storage capacity of exfoliated turbostratic carbon nanofibers. *International Journal of Hydrogen Energy* **2010**, *35*(9), 4123–4130.

25. Jimenez, V., Ramirez-Lucas, A., Sanchez, P., Valverde, J.L., Romero, A. Improving hydrogen storage in modified carbon materials. *International Journal of Hydrogen Energy* **2012**, *37*(5), 4144–4160.

26. Tsao, C.S., Tzeng, Y.R., Yu, M.S., Wang, C.Y., Tseng, H.H., Chung, T.Y., Wu, H.C., Yamamoto, T., Kaneko, K., Chen, S.H. Effect of catalyst size on hydrogen storage capacity of Pt-impregnated active carbon via spillover. *Journal of Physical Chemistry Letters* **2010**, *1*(7), 1060–1063.

27. Noh, J.S., Agarwal, R.K., Schwarz, J.A. Hydrogen storage-systems using activated carbon. *International Journal of Hydrogen Energy* **1987**, *12*(10), 693–700.

28. Im, J.S., Park, S.J., Lee, Y.S. Superior prospect of chemically activated electrospun carbon fibers for hydrogen storage. *Materials Research Bulletin* **2009**, *44*(9), 1871–1878.

29. Sevilla, M., Mokaya, R., Fuertes, A.B. Ultrahigh surface area polypyrrole-based carbons with superior performance for hydrogen storage. *Energy & Environmental Science* **2011**, *4*(8), 2930–2936.

30. Zhao, W., Fierro, V., Zlotea, C., Izquierdo, M.T., Chevalier-Cesar, C., Latroche, M. Celzard, A., Activated carbons doped with Pd nanoparticles for hydrogen storage. *International Journal of Hydrogen Energy* **2012**, *37*(6), 5072–5080.

31. Yuan, S.W., Kirklin, S., Dorney, B., Liu, D.J., Yu, L.P. Nanoporous polymers containing stereocontorted cores for hydrogen storage. *Macromolecules* **2009**, *42*(5), 1554–1559.

32. Xiang, Z.H., Cao, D.P., Wang, W.C., Yang, W.T., Han, B.Y., Lu, J.M. Postsynthetic lithium modification of covalent-organic polymers for enhancing hydrogen and carbon dioxide storage. *Journal of Physical Chemistry C* **2012**, *116*(9), 5974–5980.

33. Figueroa-Torres, M.Z., Dominguez-Rios, C., Cabanas-Moreno, J.G., Vega-Becerra, O., Aguilar-Elguezabal, A. The synthesis of Ni-activated carbon nanocomposites via electroless deposition without a surface pretreatment as potential hydrogen storage materials. *International Journal of Hydrogen Energy* **2012**, *37*(14), 10743–10749.

34. Zhevago, N.K., Glebov, V.I. Hydrogen storage in capillary arrays. *Energy Conversion and Management* **2007**, *48*(5), 1554–1559.

35. Zhevago, N.K., Denisov, E.I., Glebov, V.I. Experimental investigation of hydrogen storage in capillary arrays. *International Journal of Hydrogen Energy* **2010**, *35*(1), 169–175.

36. Holtappels, K., Beckmann-Kluge, M., Gebauer, M., Eliezer, D. Pressure Resistance of Glass Capillaries for Hydrogen Storage. *Materials Testing* **2011**, *53*(1–2), 14–18.

37. Schmitt, M.L., Shelby, J.E., Hall, M.M. Preparation of hollow glass microspheres from sol-gel derived glass for application in hydrogen gas storage. *Journal of Non-Crystalline Solids* **2006**, *352*(6–7), 626–631.

38. Qi, X.B., Gao, C., Zhang, Z.W., Chen, S.F., Li, B., Wei, S. Production and characterization of hollow glass microspheres with high diffusivity for hydrogen storage. *International Journal of Hydrogen Energy* **2012**, *37*(2), 1518–1530.

39. Weitkamp, J., Fritz, M., Ernst, S. Zeolites as media for hydrogen storage. *International Journal of Hydrogen Energy* **1995**, *20*(12), 967–970.

40. Langmi, H.W., Book, D., Walton, A., Johnson, S.R., Al-Mamouri, M.M., Speight, J.D., Edwards, P.P., Harris, I.R., Anderson, P.A. Hydrogen storage in ion-exchanged zeolites. *Journal of Alloys and Compounds* **2005**, *404*, 637–642.

41. Dong, J.X., Wang, X.Y., Xu, H., Zhao, Q., Li, J.P. Hydrogen storage in several microporous zeolites. *International Journal of Hydrogen Energy* **2007**, *32*(18), 4998–5004.

42. Chung, K.H. High-pressure hydrogen storage on microporous zeolites with varying pore properties. *Energy* **2010**, *35*(5), 2235–2241.

43. Yang, G., Zhou, L.J., Liu, X.C., Han, X.W., Bao, X.H. Adsorption, reduction and storage of hydrogen within ZSM-5 zeolite exchanged with various ions: A comparative theoretical study. *Microporous and Mesoporous Materials* **2012**, *161*, 168–178.

44. Rosi, N.L., Eckert, J., Eddaoudi, M., Vodak, D.T., Kim, J., O'Keeffe, M., Yaghi, O.M. Hydrogen storage in microporous metal-organic frameworks. *Science* **2003**, *300*(5622), 1127–1129.

45. Wong-Foy, A.G., Matzger, A.J., Yaghi, O.M. Exceptional H-2 saturation uptake in microporous metal-organic frameworks. *Journal of the American Chemical Society* **2006**, *128*(11), 3494–3495.

46. Yan, Y., Lin, X., Yang, S.H., Blake, A.J., Dailly, A., Champness, N.R., Hubberstey, P., Schroder, M. Exceptionally high H-2 storage by a metal-organic polyhedral framework. *Chemical Communications* **2009**, (9), 1025–1027.

47. Yang, S.J., Jung, H., Kim, T., Im, J.H., Park, C.R. Effects of structural modifications on the hydrogen storage capacity of MOF-5. *International Journal of Hydrogen Energy* **2012**, *37*(7), 5777–5783.

48. Yang, S.J., Kim, T., Im, J.H., Kim, Y.S., Lee, K., Jung, H., Park, C.R. MOF-Derived hierarchically porous carbon with exceptional porosity and hydrogen storage capacity. *Chemistry of Materials* **2012**, *24*(3), 464–470.

49. Mao, W.L., Mao, H.K. Hydrogen storage in molecular compounds. *Proceedings of the National Academy of Sciences of the United States of America* **2004**, *101*(3), 708–710.

50. Lee, H., Lee, J.W., Kim, D.Y., Park, J., Seo, Y.T., Zeng, H., Moudrakovski, I.L., Ratcliffe, C.I., Ripmeester, J.A. Tuning clathrate hydrates for hydrogen storage. *Nature* **2005**, *434*(7034), 743–746.

51. Su, F.B., Bray, C.L., Carter, B.O., Overend, G., Cropper, C., Iggo, J.A., Khimyak, Y.Z., Fogg, A.M., Cooper, A.I. Reversible hydrogen storage in hydrogel clathrate hydrates. *Advanced Materials* **2009**, *21*(23), 2382–2386.

52. Clawson, J.S., Cygan, R.T., Alam, T.M., Leung, K., Rempe, S.B. Ab initio study of hydrogen storage in water clathrates. *Journal of Computational and Theoretical Nanoscience* **2010**, *7*(12), 2602–2606.

53. Belosludov, R.V., Zhdanov, R.K., Subbotin, O.S., Mizuseki, H., Souissi, M., Kawazoe, Y., Belosludov, V.R. Theoretical modelling of the phase diagrams of clathrate hydrates for hydrogen storage applications. *Molecular Simulation* **2012**, *38*(10), 773–780.

54. Willow, S.Y., Xantheas, S.S. Enhancement of hydrogen storage capacity in hydrate lattices. *Chemical Physics Letters* **2012**, *525–526*, 13–18.

8

Hydrogen Utilization: Combustion

8.1 BASICS ABOUT COMBUSTION

Combustion of fuel such as hydrogen or hydrocarbons in air or oxygen is an important process for many applications, including combustion engines. The chemistry of combustion is actually very complex, involving many elementary reactions and radial species, often called radical pools. The reaction mechanisms are highly dependent on many factors, such as temperature, pressure, reactor, catalyst, and composition.

Hydrogen gas is highly flammable and can burn in air at a very wide range of concentrations between 4% and 75% by volume. The enthalpy of combustion for hydrogen is -286 kJ·mol^{-1}:

$$H_2(g) + \frac{1}{2} O_2(g) \rightarrow H_2O(l) + 286 \text{ kJ}. \tag{8.1}$$

Hydrogen mixed with air in the concentration range of 4–75% can spontaneously explode by spark, heat, or sunlight, and the hydrogen autoignition temperature, the temperature of spontaneous ignition in air, is 500°C or 932°F. Pure hydrogen–oxygen flames emit ultraviolet light and are essentially invisible to the naked eye, which presents a potential safety hazard. Hydrogen also reacts with other oxidizing elements. For example, it can react

Hydrogen Generation, Storage, and Utilization, First Edition. Jin Zhong Zhang, Jinghong Li, Yat Li, and Yiping Zhao.
© 2014 John Wiley & Sons, Inc. Published 2014 by John Wiley & Sons, Inc.

139

spontaneously and violently at room temperature with chlorine and fluorine to form the corresponding hydrogen halides, which are useful and dangerous acids.

8.2 MECHANISM OF COMBUSTION

Detailed flow reactor studies and kinetic modeling of the H_2/O_2 reaction has been reviewed and updated by Dryer and co-workers. This section will present a brief summary based on the review of Dryer et al. [1, 2]. The updated H_2/O_2 reaction mechanism consists of 19 reversible elementary reactions summarized in Table 8.1. These reactions can be categorized into four

TABLE 8.1   **Summary of 19 Reversible Elementary Reactions in the H_2O_2 Reaction Mechanism**

H_2/O_2 chain reactions
1. $H + O_2 = O + OH$
2. $O + H_2 = H + OH$
3. $H_2 + OH = H_2O + H$
4. $O + H_2O = OH + OH$

H_2/O_2 dissociation/recombination reactions
5. $H_2 + M = H + H + M$
    $H_2 + Ar = H + H + M$
6. $O + O + M = O_2 + M$
    $O + O + Ar = O_2 + Ar$
7. $O + H + M = OH + M$
8. $H + OH + M = H_2O + M$
    $H + OH + Ar = H_2O + Ar$

Formation and consumption of HO_2
9. $H + O_2 + M = HO_2 + M$
    $H + O_2 + Ar = HO_2 + M$
    $H + O_2 = HO_2$
10. $HO_2 + H = H_2 + O_2$
11. $HO_2 + H = OH + OH$
12. $HO_2 + O = OH + O_2$
13. $HO_2 + OH = H_2O + O_2$

Formation and consumption of H_2O_2
14. $HO_2 + HO_2 = H_2O_2 + O_2$
    $HO_2 + Ar = OH + OH + Ar$
15. $H_2O_2 + M = OH + OH + M$
    $H_2O_2 + Ar = OH + OH + Ar$
    $H_2O_2 = OH + OH$
16. $H_2O_2 + H = H_2O + OH$
17. $H_2O_2 + H = H_2 + HO_2$
18. $H_2O_2 + O = OH + HO_2$
19. $H_2O_2 + OH = H_2O + HO_2$
    $H_2O_2 + OH = H_2O + HO_2$

Note: M for nonreactive gas such as N_2, Ar, or He.

groups: (1) H_2/O_2 chain reactions, (2) H_2/O_2 dissociation/recombination reactions, (3) formation and consumption of HO_2, and (4) formation and consumption of H_2O_2.

Thermodynamics data and rate coefficients for these reactions are given in Mueller et al. [1] and updated in Li et al. [2] for reactions 1 to 3. Most of the early profile measurements of the H_2O_2 reaction have been obtained using a variable pressure flow reactor (VPFR) over pressure ranges of 0.3–15.7 atm and temperature range of 850–1040 K, respectively [1]. These data span the explosion limit behavior of the system and place significant emphasis on HO_2 and H_2O_2 kinetics. The reaction mechanism proposed was later updated with newer experimental results and validated against a wide range of experimental conditions, including those in shock tubes, flow reactors, and laminar premixed flames.[2] Overall, excellent agreement was found between the model predictions and experimental observations, demonstrating good predictive capabilities of the proposed reaction mechanism for different experimental systems. Good agreement between different models has also been found. For example, Figure 8.1 shows the temperature and pressure dependence of the rate constant of the reaction $H + O_2 (+M) \rightarrow HO_2 (+M)$ (for $M = N_2$) predicted by two different models [2, 3], which agree reasonably well (within 20%) with each other over 300–3000 K and from low to high pressure range.

The reaction rate constant for each of the reaction can be expressed in Arrhenius form as follows:

$$k = AT^n \exp\left(-\frac{E_a}{RT}\right),\tag{8.2}$$

where A is a prefactor, T is temperature (K), n is a number between -2.00 and 2.67 depending the specific reaction and is zero for many of the reactions, especially reactions involving the formation and consumption of HO_2 and H_2O_2, E_a is the activation energy, and R is the gas constant. The different parameters and related thermodynamics data have been obtained from modeling of experimental results.[1, 2] For example, the low-pressure-limit rate constant for the chain termination reaction ($H + O_2 + M = HO_2 + M$) for $M=N_2$ or Ar has been found to be (in unit of $cm^6 \, mol^{-2} \, s^{-1}$):

$$k_0^{N_2} = 6.37 \times 10^{20} T^{-1.72} \exp\left(-\frac{264}{T}\right)\tag{8.3}$$

$$k_0^{Ar} = 9.04 \times 10^{19} T^{-1.50} \exp\left(-\frac{248}{T}\right).\tag{8.4}$$

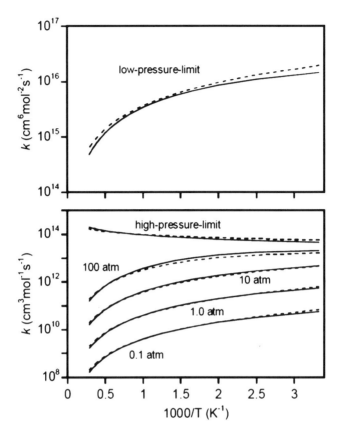

FIGURE 8.1   Temperature and pressure dependence of the reaction rate of $H + O_2 (+M) \rightarrow HO_2 (+M)$ for $M = N_2$. Solid lines represent the values used in the mechanism proposed in Li et al. [2], and dashed lines the recommendations of Troe [3]. *Source*: Reproduced with permission from Li et al. [2].

The parameters are clearly dependent on the specific third-body gas used.

Among the reactions, the $H + OH + M$ reaction was found to be primarily important only to laminar flame speed propagation predictions at high pressure. The overall H_2/O_2 reaction system is very sensitive to the key chain branching reaction ($H + O_2 = O + OH$) and the important chain termination reaction ($H + O_2 + M = HO_2 + M$). The branching ratio of the rate constants of these two reactions depends strongly on temperature and pressure, and results from an earlier and subsequently updated models are compared and shown in Figure 8.2 [1, 2]. The agreement is excellent in the temperature range of 800–900 K, where the value of rate constant used in the earlier model by Mueller et al. was experimentally derived.

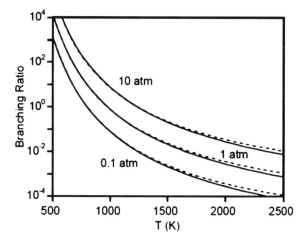

FIGURE 8.2   Branching ratio of rate constants of the chain branching reaction ($H + O_2 = O + OH$) and the chain termination reaction ($H + O_2 + M = HO_2 + M$) as a function of temperature for three different pressures based on two different models: solid lines [2] and dashed lines [1]. *Source*: Reproduced with permission from Li et al. [2].

8.3   MAJOR FACTORS AFFECTING COMBUSTION

One of the most important factors affecting H_2 combustion is presence of other gases such as O_2 and CH_4 that participate in combustion or other gases such as N_2 or Ar that affect combustion. The presence of O_2 is critical for combustion, and the ratio between H_2 and O_2 is thus a key parameter. Mixing with other active gases such as CH_4 has major consequences on the outcome and mechanism of combustion, including emission products. For example, mixtures of hydrogen and methane as fuel allow a substantial reduction of CO_2 emissions at affordable costs [4].

When hydrogen combustion is carried in air or with N_2 present, nitrogen species, particularly NO_x (x indicating various ratios between N and O), play a critical role in the combustion process [5]. Several important reactions involving NO_x are summarized in Table 8.2.

The two main reaction paths forming NO_x relevant in H_2 combustion are thermal NO (reactions 3–5 in Table 8.2) and the nitrous oxide mechanism (N_2O) (reaction 1). The NNH mechanism (reaction 2) is always important at low temperatures while only relevant at low residence times for rich mixtures at high temperature. Figure 8.3 show a dependence of maximum [NO] on maximum temperature [6]. It is clear that the overall maximum [NO] increases with the maximum temperature in the range of about 1850–2150 K.

TABLE 8.2   Summary of Several Important Reactions Involving NO$_x$ in Hydrogen Combustion

NO formation
    1. $N_2O + O = NO + NO$
    2. $NNH + O = NH + NO$
    3. $N + O_2 = NO + O$
    4. $N + N_2 = NO + N$
    5. $N + OH = NO + H$
NO removal
    6. $NO + HO_2 = NO_2 + OH$
    7. $H + NO + M = HNO + M$
    8. $NH + NO = N_2O + H$
    9. $N_2H_2 + NO = N_2O + NH_2$
Other reactions
    10. $N_2O + M = N_2 + O + M$
    11. $N_2O + H = N_2 + OH$
    12. $N_2O + O = N_2 + O_2$
    13. $NO_2 + H_2 = HONO + H$

Note: M for nonreactive gas such as N_2, Ar, or He.

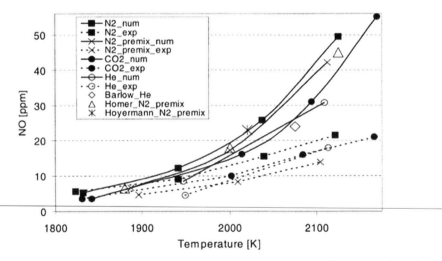

FIGURE 8.3 Dependence of maximum [NO] on maximum temperature for different experimental results and numerical simulations [6].

Dilution with gases such as N_2 reduces the flame temperature and thus the thermal mechanism of NO$_x$ formation. At high dilutions, the thermal mechanism is almost suppressed and the NO$_x$ formation occurs mainly through NNH and N_2O mechanisms [5].

NO$_x$ emission from hydrocarbon combustion is strongly influenced by hydrogen addition. For example, in a recent study of mild flameless combustion regime applied to methane/hydrogen mixtures in a laboratory-scale pilot furnace with or without air preheating, results show that mild flameless combustion regime is achieved from pure methane to pure hydrogen whatever the CH_4/H_2 proportion [7]. The main reaction zone remains lifted from the burner exit, in the mixing layer of fuel and air jets, ensuring a large dilution correlated to low NO_x emissions, whereas CO_2 concentrations obviously decrease with hydrogen proportion. A decrease of NO_x emissions is measured for larger quantity of hydrogen due mainly to the decrease of prompt NO formation. Without air preheating, a slight increase of the excess air ratio is required to control CO emissions. For pure hydrogen fuel without air preheating, mild flameless combustion regime leads to operating conditions close to a "zero emission furnace," with ultralow NO_x emissions and without any carbonated species emissions.

Another important factor in hydrogen combustion is pressure. For example, a chemical kinetic model for high pressure combustion of H_2/O_2 mixtures has been developed recently by updating some of the rate constants important under high pressure conditions without any diluents [8]. The revised mechanism is validated against experimental shock-tube ignition delay times and laminar flame speeds, with predictions of the model compared with those by several other kinetic models. While predictions of the different models agree well with each other and with the experimental data of ignition delay times and flame speeds at pressures lower than 10 atm, substantial differences are observed between experimental data of high pressure mass burning rates and model predictions, as well as among the model predictions themselves. Different pressure dependencies of mass burning rates above 10 atm in different kinetic models result from using different rate constants in these models for HO_2 reactions, especially for $H + HO_2$ and $OH + HO_2$ reactions. The rate constants for the reaction $H + HO_2$ involving different product pathways were found to be very important for predicting high pressure combustion properties.

Temperature is another important factor in hydrogen combustion. In particular, high temperature hydrogen combustion is of interest to nuclear reactor safety [9]. In a combined experimental and simulation study, it has been found that the mass flux or burning velocity increases exponentially with increasing temperature in a wide temperature range, as shown in Figure 8.4 [10]. In this example, the experimentally measured temperature was based on coherent anti-Stokes Raman scattering (CARS). The calculations were carried out using different standard models. By comparing the computed variation of flame temperature with mass flux in burner-stabilized flat flames with those obtained experimentally, the predictive power of a chemical mechanism was tested at constant equivalence ratio over a range of more

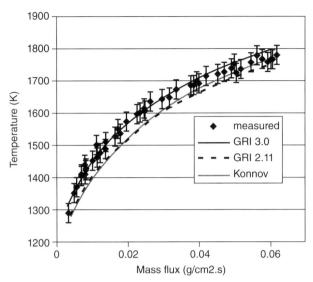

FIGURE 8.4   Flame temperature vs. mass flux at equivalence ratio $\varphi = 0.6$. The equivalence ratio is defined as the ratio of the actual fuel/air ratio to the stoichiometric fuel/air ratio. Points: measurements; lines: calculations using different models or reaction mechanisms: GRI-Mech 3.0 (thick solid line) or GRI-Mech 2.11 (dashed line) or Konnov (thin solid line). *Source*: Reproduced with permission from Haruta and Sano [10].

than 700 K. The approach was suggested to be general and can be used for other fuels as well.

8.4   CATALYTIC COMBUSTION

Hydrogen combustion can be assisted by catalysis as in many chemical reactions. Compared with flame-type combustion, catalytic combustion can operate at lower temperature and reduce pollution due to NO_x emission in flame combustion [11]. Common catalysts used include Pt, Pd, Ni, and Co, usually dispersed in metal oxide supports, for example, Al_2O_3, SiO_2, ZrO_2, and TiO_2. One of the primary functions of the catalysts is to facilitate dissociation of the H_2 molecule by weakening the H–H bond through adsorption onto the catalyst surface. The supports help to disperse the metal catalysts and enhance surface area, while in some cases, play a more active role by functioning as catalysts themselves. For example, an earlier experimental study has examined the effect of fundamental conditions of hydrogen-fueled catalytic burners on their operating properties [12]. In diffusive combustion, Pd-power coated Ni foam with relatively large pores could afford the highest

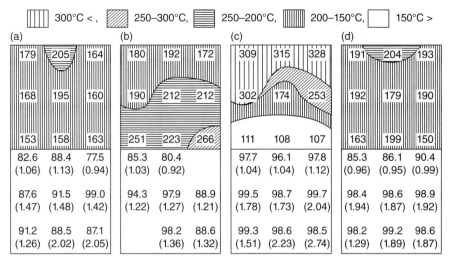

FIGURE 8.5 Distribution of spot temperature, combustion efficiency (E) and equivalent air ratio (λ_2) over catalyst surface in diffusive combustion. Heat input, 1.2 kcal·cm^{-2}·h^{-1}. Equivalent ratio of secondary air, $\lambda_2 = 1.8\ 0.2$. (a) Catalyst T$_h$, $\lambda_2 = 1.74$, E $= 83.7 - 87.3\%$. (b) Catalyst D$_5$, $\lambda_2 = 1.62$, $E = 81.2\%$. (c) Catalyst D$_2$, $\lambda_2 = 1.96$, $E = 98.0\%$. (d) Catalyst G$_0$, $\lambda_2=1.94$, $E = 93.0\%$. *Source*: Reproduced with permission from Haruta et al. [12].

combustion efficiency; however, it suffered from considerable nonuniform distribution of surface temperature. It was demonstrated that a ceramic foam coated with Co–Mn–Ag oxide was practically utilizable for catalytic appliances operating on hydrogen fuel, although it required a little preheating for initiating combustion. Premixing of air to a 40% stoichiometric amount with hydrogen was effective in improving combustion efficiency. In a completely premixed operation, combustion efficiency higher than 99% was obtained in the heat input range up to 1.6 kcal·cm^{-2}·h^{-1}. Figure 8.5 shows representative distribution of spot tempearature, combustion efficiency (E), and equivalent air ratio over catalyst surface in diffusive combustion for several different catalysts studied. The combustion efficiency is sensitive to both temperature of the catalyst surface and the equivalent ratio.

In a more recent study, a mesoporous ceramic coating monolithic Pt-based catalyst (Pt/Ce$_{0.6}$Zr$_{0.4}$O$_2$/MgAl$_2$O$_4$/cordierite) was prepared from inorganic salt and alkali and used for catalytic hydrogen combustion study [13]. The results showed that the coating had typical spinel crystallization structure and high surface area (more than 200 m^2·g^{-1}). The Ce-Zr oxide modification increased the surface area and improved the oxygen storage capacity. And the results of hydrogen catalytic combustion indicated that this monolithic catalyst had high activity for hydrogen combustion reaction, which could quickly start up even at 263 K. For low temperature catalytic combustion of

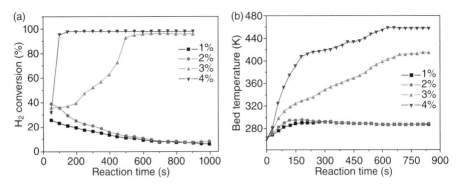

FIGURE 8.6 (a) H_2 conversion rates and (b) bed temperature changes versus reaction time over the Pt/$Ce_{0.6}Zr_{0.4}O_2$/$MgAl_2O_4$ cordierite catalyst under different hydrogen concentrations at 263 K as initial temperature (20,000 h^{-1}). *Source*: Reproduced with permission from Zhang et al. [13].

hydrogen, the initial reaction temperature, H_2 concentration, and space velocity were very important parameters. Figure 8.6 shows the H_2 conversion rates and bed temperature changes as a function of reaction time over the Pt/ $Ce_{0.6}Zr_{0.4}O_2$/$MgAl_2O_4$ cordierite catalyst under different hydrogen concentrations. The H_2 conversion rate is very sensitive to the hydrogen concentration, and changes substantially from 2% to 3%.

The supports of catalysts can also have important influence on the hydrogen combustion reaction. For example, a recent study has compared the effect of several different oxide and vanadate compounds (CeO_2, Fe_2O_3, $CeVO_4$, and $FeVO_4$) as support on catalytic hydrogen combustion [14]. While all the four compounds showed good activity and stability for catalytic hydrogen combustion and more than 95% conversion was observed over all the compounds in the temperature range of 250–500°C, the reaction mechanisms differ for the different compounds in the nature of adsorption of H_2 over the adsorption sites. The rates of reaction were very high when the reaction was carried out with high concentrations of O_2 in the stream. $FeVO_4$ was found to be the best catalyst, and the possible reason behind the high activity of the compound was the presence of favorable Fe^{3+}–Fe^{2+} and V^{5+}–V^{4+} redox couples. The reaction was found to proceed via dissociative H_2 adsorption mechanism over ceria-based compounds while it was found to proceed via molecular adsorption over iron-based catalysts. Figure 8.7 shows the proposed reaction mechanism over Fe-based compounds. The middle panel shows molecularly adsorbed H_2 in the compound abstracting oxygen from the lattice, resulting in the release of H_2O and reduced iron oxide with vacancies. The bottom panel illustrates the dissociation of O_2 over the oxygen ion vacancies, reoxidation of the support, formation of weakly adsorbed

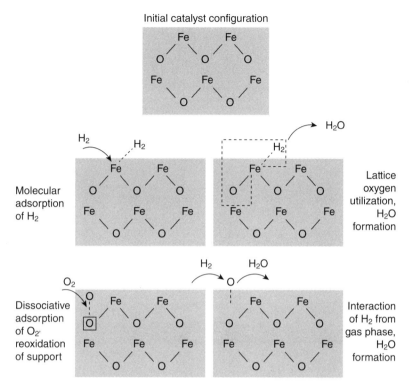

FIGURE 8.7 Proposed mechanism of catalytic hydrogen combustion over Fe-based compound. *Source*: Reproduced with permission from Deshpande et al. [14].

oxygen species on the support, interaction of H_2 from gas phase with oxygen adsorbed in the support, and release of H_2O giving back the catalyst in its original form.

8.5 SUMMARY

Hydrogen combustion is a fast, efficient, and complex process involving many elementary reactions. The reaction mechanisms are highly dependent on many factors, including pressure, temperature, reactor structure, and carrier or dilute gases. The reaction is extremely useful and can be dangerous. Thus, major precaution is required in dealing with the reaction. While substantial research has been done on this reaction, the detailed reaction mechanism still could use further research, partly due to its complex and fast nature. With advancement in computational chemistry, new insight into the reaction mechanism could be gained from the theoretical and computational front, especially given that most of the elements involved in the combustion reaction are light, rendering themselves convenient for theoretical studies. On the experimental front, ultrafast kinetic studies may help to gain more and direct insight into the reaction mechanism involving short-lived species, such as radicals as intermediates.

REFERENCES

1. Mueller, M.A., Kim, T.J., Yetter, R.A., Dryer, F.L. Flo reactor studies and kinetic modeling of the H_2/O_2 reaction. *Int. J. Chem. Kinetics* **1999**, *31*, 113–125.

2. Li, J., Zhao, Z.W., Kazakov, A., Dryer, F.L. An updated comprehensive kinetic model of hydrogen combustion. *International Journal of Chemical Kinetics* **2004**, *36*(10), 566–575.

3. Troe, J., Detailed modeling of the temperature and pressure dependence of the reaction $H+O_2$ (+M) → HO_2 (+M). *Proc. Combust. Inst* **2000**, *28*, 1463–1469.

4. Klell, M., Eichlseder, H., Sartory, M. Mixtures of hydrogen and methane in the internal combustion engine: Synergies, potential and regulations. *International Journal of Hydrogen Energy* **2012**, *37*(15), 11531–11540.

5. Frassoldati, A., Faravelli, T., Ranzi, E. A wide range modeling study of NOx formation and nitrogen chemistry in hydrogen combustion. *International Journal of Hydrogen Energy* **2006**, *31*(15), 2310–2328.

6. Rortveit, G.J., Hustad, J.E., Li, S.C., Williams, F.A. Effects of diluents on NOx formation in hydrogen counterflow flames. *Combustion and Flame* **2002**, *130*(1–2), 48–61.

7. Ayoub, M., Rottier, C., Carpentier, S., Villermaux, C., Boukhalfa, A.M., Honore, D. An experimental study of mild flameless combustion of methane/hydrogen mixtures. *International Journal of Hydrogen Energy* **2012**, *37*(8), 6912–6921.

8. Shimizu, K., Hibi, A., Koshi, M., Morii, Y., Tsuboi, N. Updated kinetic mechanism for high-pressure hydrogen combustion. *Journal of Propulsion and Power* **2011**, *27*(2), 383–395.

9. Stamps, D.W., Berman, M. High-temperature hydrogen combustion in reactor safety applications. *Nuclear Science and Engineering* **1991**, *109*(1), 39–48.

10. Sepman, A.V., Mokhov, A.V., Levinsky, H.B. Extending the predictions of chemical mechanisms for hydrogen combustion: Comparison of predicted and measured flame temperatures in burner-stabilized, 1-D flames. *International Journal of Hydrogen Energy* **2011**, *36*(15), 9298–9303.

11. Haruta, M., Sano, H. Catalytic combustion of hydrogen 1: Its role in hydrogen utilization system and screening of catalyst materials. *International Journal of Hydrogen Energy* **1981**, *6*(6), 601–608.

12. Haruta, M., Souma, Y., Sano, H. Catalytic combustion of hydrogen 2: An experimental investigation of fundamental conditions for burner design. *International Journal of Hydrogen Energy* **1982**, *7*(9), 729–736.

13. Zhang, C.M., Zhang, J., Ma, J.X. Hydrogen catalytic combustion over a Pt/Ce0.6Zr0.4O2/ MgAl$_2$O$_4$ mesoporous coating monolithic catalyst. *International Journal of Hydrogen Energy* **2012**, *37*(17), 12941–12946.

14. Deshpande, P.A., Polisetti, S., Madras, G., Analysis of oxide and vanadate supports for catalytic hydrogen combustion: Kinetic and mechanistic investigations. *Aiche Journal* **2012**, *58*(3), 932–945.

9

Hydrogen Utilization: Fuel Cells

9.1 BASICS OF FUEL CELLS

9.1.1 The Rational Development of Fuel Cells

Hydrogen has the highest energy content by weight (33,320 Wh·kg^{-1}), which is about 2.6 and 2.4 times more than gasoline (12,700 Wh·kg^{-1}) and natural gas (13,900 Wh·kg^{-1}), respectively [1–3]. In Chapter 8, we have reviewed hydrogen utilization for direct combustion. However, the efficiency of a hydrogen internal combustion engine (ICE) is ca. 25% while that of a hydrogen fuel cell vehicle is 60%, which is three times better than that of petrol-fuelled engines (18–20% for a petrol ICE reaching 40% at peak efficiency) [4, 5].

In this chapter, we will discuss the application of hydrogen in fuel cells. A fuel cell is a device that converts chemical energy supplied as input fuels into electric energy, and is essentially a battery with an external fuel source [5]. It creates electricity by stripping electrons off the hydrogen. Like a battery, a typical fuel cell contains a set of plates for electrochemical reactions [6]. Unlike a battery, it never runs out and it produces a continuous flow of clean power as long as fuel is supplied. The invention of fuel cells as energy conversion systems dates back from the middle of the nineteenth century by Sir William Grove, "father of the fuel cell." However, the principle

Hydrogen Generation, Storage, and Utilization, First Edition. Jin Zhong Zhang, Jinghong Li, Yat Li, and Yiping Zhao.
© 2014 John Wiley & Sons, Inc. Published 2014 by John Wiley & Sons, Inc.

was first discovered by Christian Friedrich Schönbein [3]. Faced with major environmental issues in the use of fossil fuels for applications such as electricity generation and automobiles, hydrogen fuel cells provide an attractive alternative due to its high efficiency and clean byproduct (water) [7]. Significantly increased efforts have been made recently to advance the fuel cell technology and understanding of related fundamental issues.

9.1.2 Work Principles of Fuel Cells

A fuel cell produces electricity from electrochemical oxidation of the fuel. An electrochemical cell usually consists of two electrodes that allow the overall reaction described below to take place:

$$A_{ox1} + B_{red1} \rightarrow C_{red2} + D_{ox2}, \tag{9.1}$$

where A_{ox1} is the oxidant, usually O_2 in the fuel cells, B_{red1} is the reducer, usually H_2 or other hydrocarbon fuels, C_{red2} is the reductive, usually H_2O in fuel cells, and D_{ox2} is usually CO_2 when the fuel is hydrocarbon. The Gibbs free energy change of a chemical reaction is related to the cell voltage via:

$$\Delta G = -nF\Delta U_0, \tag{9.2}$$

where n is the number of electrons involved in the reaction, F is the Faraday constant, and ΔU_0 is the voltage of the cell for thermodynamic equilibrium in the absence of a current flow.

There are many types of fuel cells, but they all consist of an anode (negative side), a cathode (positive side), and an electrolyte that allows charges to move between the two electrodes of the fuel cell [7]. Electrons are drawn from the anode to the cathode through an external circuit, producing direct current electricity. Fuel cells come in a variety of sizes. Individual fuel cells produce relatively small electrical potentials, about 0.7 V, so cells are "stacked" or placed in series to increase the voltage and meet specific application requirements.

Figure 9.1 shows a schematic of a typical hydrogen/oxygen fuel cell and its reactions based on the proton exchange membrane [4]. This fuel cell is constructed using polymer electrolyte membranes (notably Nafion) as proton conductor and platinum (Pt)-based materials as catalyst. Its noteworthy features include low operating temperature, high power density, and easy scale-up, making it a promising candidate as the next-generation power sources for transportation, stationary, and portable applications. It is manufactured as a stack of identical repeating unit cells comprising a membrane

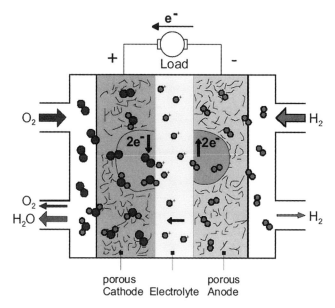

FIGURE 9.1   Schematic drawing of a hydrogen/oxygen fuel cell and its reactions based on the proton exchange membrane fuel cell. *Source*: Reproduced with permission from Carrette et al. [4]. (See color insert.)

electrode assembly (MEA) in which hydrogen gas (H_2) is oxidized on the anode and oxygen gas (O_2) is reduced on the MEA cathode, all compressed by bipolar plates that introduce gaseous reactants and coolants to the MEA and harvest the electric current. The electrochemical reactions occur at the MEA electrodes, each attached to a solid polymer ion exchange membrane that conducts protons but not electrons. Both the cathode oxygen reduction reaction (ORR) and anode hydrogen oxidation reaction occur on the surfaces of Pt-based catalysts. Pure water and heat are the only by-products. Porous gas diffusion layers transport reactants and product water between the flow fields and catalyst surfaces while exchanging electrons between them. The major reactions are given below.

$$\text{Anode: } H_2 \rightarrow 2H^+ + 2e^- \tag{9.3}$$

$$\text{Cathode: } 4H^+ + 4e^- + O_2 \rightarrow 2H_2O \tag{9.4}$$

$$\text{Overall: } 2H_2 + O_2 \rightarrow 2H_2O. \tag{9.5}$$

9.1.3   Operation of Fuel Cells

The operation of fuel cells is virtually electrochemical reactions, and the overall properties can be characterized by Nernst equation. For the oxidation of hydrogen, the kinetics of this reaction is very fast on Pt catalysts, and in a fuel cell, the oxidation of hydrogen at higher current densities is usually controlled by mass-transfer limitations. The oxidation of hydrogen also involves the adsorption of the gas onto the catalyst surface followed by dissociation of the molecule and electrochemical reaction to two hydrogen ions as follows:

$$2Pt\ (s) + H_2 \rightarrow Pt-H_{ads} + Pt-H_{ads} \qquad (9.6)$$

$$Pt-H_{ads} \rightarrow H^+ + e^- + Pt_{(s)}, \qquad (9.7)$$

where $Pt_{(s)}$ is a free surface site and $Pt-H_{ads}$ is an adsorbed H atom on the Pt active site. The overall reaction of hydrogen oxidation is:

$$H_2 \rightarrow 2H^+ + 2e^- \quad U^0 = 0\ V. \qquad (9.8)$$

The electrocatalytic oxygen reduction reaction (ORR) on catalyst surfaces [RHE = reversible hydrogen electrode] is one of the most widely studied reactions in electrochemistry. Its fundamental and technological importance is based on the fact that the oxygen/water half-cell reaction is a strongly oxidizing and ubiquitous redox couple [7]. Combined with an electron-supplying redox process, a direct electrochemical conversion of hydrogen, the overall Gibbs energy of reaction into electrical potentials is achieved. This conversion is the scientific basis for electrochemical reactions in fuel cells or metal–air batteries. The ORR is also used in oxygen depolarization cathodes (ODC) in modern chlorine technologies, in which it replaces the hydrogen evolution process to improve electrical efficiencies. The reverse ORR process, that is, the evolution of oxygen from water, is crucial for efficient water (photo)-electrolysis into hydrogen or in metal electrodeposition processes in the semiconductor industry [4, 5].

The efficiency of a fuel cell can be calculated from the Gibbs free energy (ΔG) and the enthalpy change (ΔH) of the electrochemical reaction. Ideally, the free energy of the reaction can be completely converted into electrical energy and the efficiency is given by:

$$\varepsilon = \frac{W_e}{(-\Delta H)} = \frac{nF\Delta Uo}{(-\Delta H)} = \frac{\Delta G}{\Delta H} = 1 - \frac{T\Delta S}{\Delta H}, \qquad (9.9)$$

where W_e is the electrical work performed, and ΔS is the isothermal entropy change of the reaction. $T\Delta S$ corresponds to the reversible heat exchanged with the external environment. The change in entropy of the reaction (ΔS) depends strongly on the reactants and products [4, 5].

9.2 TYPES OF FUEL CELLS

Fuel cells come in many varieties; however, they all work in the same general manner. As mentioned earlier, a typical fuel cell is made up of three adjacent segments: the anode, the electrolyte, and the cathode. Two chemical reactions occur at the interfaces of the three different segments. The net result of the two reactions is that the fuel is consumed, water or carbon dioxide is generated, and an electric current is produced, which can be used to power electrical devices, normally referred to as the load. Fuel cells are usually classified by the electrolyte employed in the cell, as summarized in Table 9.1. Direct methanol fuel cell (DMFC), with methanol directly fed to the anode, is beyond this classification, and we will mainly discuss the fuel cells based on hydrogen as fuel.

9.2.1 Alkaline Fuel Cell (AFC)

The AFC has the highest electrical efficiencies among fuel cells, but works only with the pure gases, which is a restraint for wide applications [8]. The AFC is also one of the first fuel cells used in space, and the first technological AFC was developed by the Bacon group at the University of Cambridge [9]. The initial electrodes used in AFC are Ni-based catalysts, sometimes activated with novel metals. Pt-carbon electrodes are now widely used for both the anode and cathode. The Pt-based alloys have been proved to have higher activity than bare Pt for oxygen reduction due to a higher exchange current density. Nowadays, various Pt-based alloys have been synthesized, and their sizes, morphologies, and compositions have been widely explored [10]. The alkaline KOH electrolyte is used in AFC, which has an advantage over acid fuel cells in that oxygen reduction kinetics are much faster in it than in acid electrolyte [11]. The circulating KOH also provides a good barrier against gas leakage and can be used as a cooling liquid in the cell or stack. In addition, since the high open circuit voltage of the KOH electrolyte-based AFC cell may induce the carbon oxidation and produce carbonates that are deleterious to the cell performance, circulating the KOH, instead of using a stabilized matrix, can avoid the buildup of carbonates [12].

9.2.2 Proton Exchange Membrane Fuel Cell (PEMFC)

The PEMFC (originally referred to as the solid polymer electrolyte fuel cell) was the first type of fuel cell to find an application—power source for NASA's Gemini space flights in the 1960s [13]. This technology was dormant for about 20 years thereafter. More recently, the California Environmental

TABLE 9.1 The Different Fuel Cells that Have Been Realized and Their Electrode Reactions

	AFC (Alkaline)	PEMFC (Polymer Electrolyte Membrane)	DMFC (Direct Methanol)	PAFC (Phosphoric Acid)	MCFC (Molten Carbonate)	SOFC (Solid Oxide)
Operating temp. (°C)	<100	60–120	60–120	160–220	600–800	800–1000 low temperature (500–600) possible
Anode reaction	$H_2 + 2OH^-$ $\rightarrow 2H_2O + 2e^-$	$H_2 \rightarrow 2H^+ + 2e^-$	$CH_3OH + H_2O$ $\rightarrow CO_2 + 6H^+ + 6e^-$	$H_2 \rightarrow 2H^+ + 2e^-$	$H_2 + CO_3^{2-}$ $\rightarrow H_2O + CO_2 + 2e^-$	$H_2 + O^{2-}$ $\rightarrow H_2O + 2e^-$
Cathode reaction	$\frac{1}{2}O_2 + H_2O + 2e^-$ $\rightarrow 2OH^-$	$\frac{1}{2}O_2 + 2H^+ + 2e^-$ $\rightarrow H_2O$	$3/2\,O_2 + 6H^+ + 6e^-$ $\rightarrow 3H_2O$	$\frac{1}{2}O_2 + 2H^+ + 2e^-$ $\rightarrow H_2O$	$\frac{1}{2}O_2 + CO_2 + 2e^-$ $\rightarrow CO_3^{2-}$	$\frac{1}{2}O_2 + 2e^-$ $\rightarrow O^{2-}$

Source: Reproduced with permission from Carrette et al. [5].

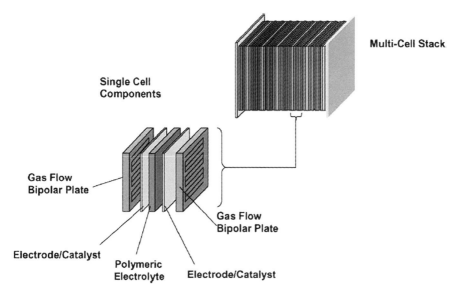

FIGURE 9.2 Schematic of PEMFC and stack. Reproduced with permission from Costamanga and Srinivasan [19].

Legislations and the U.S. Partnership for a New Generation of Vehicles program (PNGV), which was initiated in 1993 and sponsored by the U.S. government and the big three U.S. automobile manufacturers, stimulated its worldwide renaissance for the transportation application [14]. This renaissance, in turn, gave birth to the R&D programs for portable power and power generation applications [15]. Today, PEMFCs are at the forefront of the different types of fuel cells [16].

Figure 9.2 illustrates the main components of a PEMFC power source: (i) the single cell containing the porous gas diffusion electrodes (anode and cathode), the proton conducting electrolyte, anodic and cathodic catalyst layers, mostly deposited on the electrode (but more recently in some work on the proton conducting membrane), and current collectors with the reactant flow fields, (ii) a stack of cells in series, with the current collectors also serving as the bipolar plates, (iii) cell stacks (modules) connected in series or parallel, depending on the voltage and current requirements for specific applications, and (iv) auxiliaries for thermal and water management and for compression of gases [17]. The unique feature of the PEMFC, as compared with other types of fuel cells (except for the solid oxide fuel cell), is that it has a solid proton conducting electrolyte. PEMFCs operate at low temperature (below 1008 K) and generate a specific power (W·kg^{-1}) and power density (W·cm^{-2}) higher than any other type of fuel cell. It is for this reason that the PEMFC has captured attention and is the leading fuel cell candidate as power sources for transportation, small-scale power generation, and portable applications [18].

The proton conducting membrane is the vital component of a PEMFC, since that is what makes it possible to attain high power densities [7]. A major breakthrough in the field of PEMFCs came with the use of Nafion membranes by DuPont. These membranes possess a higher acidity and also a higher conductivity and are far more stable than the polystyrene sulfonate membranes. Furthermore, composite membranes have been widely explored to improve the membrane structure and conductivity. Novel membranes have also been prepared by new techniques, such as radiation grafting or plasma polymerization, which have proved to be mechanically and electrochemically stable for PEMFC applications. Moreover, membranes in PEMFC are usually filled with water to keep the conductivity high, since proton transport through a wet membrane is similar to that of an aqueous solution. Water management in the membrane is one of the major issues in PEMFC technology. One way to improve the water management is to humidify the gases coming into the fuel cell. Another form of water management can be found in the direct hydration of the membrane by mounting porous fiber wicks.

Electrodes for PEMFCs are generally porous gas diffusion electrodes to ensure the supply of reactant gases to the active zones where the noble metal catalyst is in contact with the ionic and electronic conductor [20]. Pt and Pt-based alloy are the best electrocatalysts for both hydrogen oxidation and oxygen reduction [21]. One of the major problems with the Pt electrocatalysis for hydrogen electrode is its low tolerance to CO in H_2 from reformed fuels, so the development of strategies to improve the stabilities of Pt based catalysts and search for other novel electrocatalysts have attracted wide researchers [22–24]. The basic designs for platinum catalysts are summarized in Figure 9.3, categorized by overall geometry of the catalyst and its support, and then further subdivided according to structural morphology and composition [16, 25, 26]. Comparison of the results shows that kinetic activity can change by nearly an order of magnitude when the catalyst is a discrete nanoparticle or a polycrystalline thin film, and that catalyst surface area per unit volume can affect the maximum achievable current density [27]. Also, the volume occupied by the nonactive support influences current density, while aspect ratios determine the packing of the catalyst supports and hence porosity and free-radical scavenging [28].

9.2.3 Phosphoric Acid Fuel Cell (PAFC)

PAFC is a type of fuel cell that uses liquid phosphoric acid as an electrolyte. It was developed in the mid-1960s and field-tested in the 1970s. Since then,

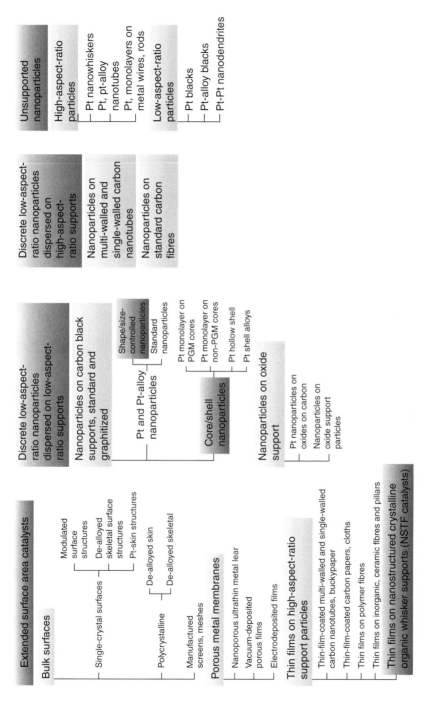

FIGURE 9.3 Basic platinum-based heterogeneous electrocatalyst approaches. *Source:* Reproduced with permission from Debe [26].

161

it has improved significantly in stability, performance, and cost [29]. These characteristics have made the PAFC a good candidate for early stationary applications [30]. In the operating range of 150–200°C, the expelled water can be converted to steam for air and water heating (combined heat and power). This potentially allows an efficiency increase of up to 70% [31]. P AFC is CO_2-tolerant ,and even can tolerate a CO concentration of about 1.5%, which broadens the choice of fuels that can be used. If gasoline is used, sulfur must be removed. At lower temperatures, phosphoric acid is a poor ionic conductor, and CO poisoning of the Pt electro-catalyst in the anode becomes severe. However, PAFC is much less sensitive to CO than PEFCs and AFCs. The disadvantage of PAFC is its rather low power density and aggressive electrolyte [30]. The electrodes used in PAFCs are generally Pt-based catalysts dispersed on a carbon-based support. For the cathode, a relatively high loading of Pt is necessary for the promotion of the O_2 reduction reaction. The hydrogen oxidation reaction at the anode occurs readily over a Pt/C catalyst. The electrolyte of PAFC is highly concentrated or pure liquid phosphoric acid (H_3PO_4) saturated in a silicon carbide matrix (SiC) [32].

9.2.4   Molten Carbonate Fuel Cell (MCFC)

The development of MCFCs started about mid-twentieth century [17]. An advantage of the MCFC is the possibility to allow for internal reforming due to the high operating temperatures (600–700°C) and to use the waste heat in combined cycle power plants [33]. A molten alkali carbonate mixture is retained in a porous lithium aluminate matrix [34]. At the cathode, oxygen reacts with carbon dioxide and electrons to form carbonate ions:

$$\tfrac{1}{2}O_2 + CO_2 + 2e^- \rightarrow CO_3^{2-}. \tag{9.10}$$

The carbonate ions flow through the electrolyte matrix from cathode to anode. At the anode, the carbonate ions are consumed by the oxidation of hydrogen to form steam and carbon dioxide, releasing electrons to the external circuit:

$$H_2 + CO_3^{2-} \rightarrow H_2O + CO_2 + 2e^-. \tag{9.11}$$

Because Ni metal anodes are not stable enough under the MCFC operating conditions, Ni–Al or Ni–Cr metal alloys have been employed as MCFC anodes [35, 36]. For cathodes, NiO is active enough for oxygen reduction at high temperatures so that a Pt-based metal is not necessary. One problem with the NiO cathode is that NiO particles grow as they creep into the molten carbonate melt, which reduces the active surface area and can short-circuit the cell. A solution to this problem is to add small amounts of Mg metal to the cathode and the electrolyte for stability [34].

The electrolyte for MCFCs is a molten carbonate that is stabilized by an alumina-based matrix [37]. Initially, Li_2CO_3/K_2CO_3 (Li/K) carbonate materials were used as electrolytes, but they tend to degrade [38]. Subsequently, an alumina phase or ceria-based materials were added to stabilize the electrolyte [4, 5].

9.2.5 Solid Oxide Fuel Cell (SOFC)

SOFCs employ a solid oxide material as electrolyte and are, thus, more stable than the molten carbonate fuel cells as no leakage problems due to a liquid electrolyte can occur [39, 40]. Due to the high operating temperature (700–1000°C), a wide variety of fuels can be processed. Anodes for SOFC are also based on Ni, and usually Ni cermet materials are used, which are more stable than plain Ni metal. A NiO powder mixed with a YSZ powder together with a resin binder produces an anode functional layer onto which the YSZ electrolyte can be deposited and sintered [41, 42].

From the beginning of SOFC development, it was found that $LaSrMnO_3$ (LSM) electrodes had a high activity for oxygen reduction at high temperatures and were stable under SOFC operation conditions [43]. These LSM cathodes have been improved over time and particularly yttria-stabilization of the cathodes improves the performance. Perovskite-type materials have been investigated as cathodes for SOFCs as well [44]. Lanthanide-based perovskites showed a high conductivity and a high catalytic activity for oxygen reduction. For electrolytes, ZrO_2- and CeO_2-based electrolytes have been found to be stable and afford reasonable conductivity.

9.3 CATALYSTS FOR OXYGEN REDUCTION REACTION OF FUEL CELLS

As we mentioned above, oxygen reduction reaction is an important half reaction in the fuel cells, and it decide the efficiency, power density durability and cost of the fuel cells [45]. In a classic PEMFC, the catalyst phase is contacted with a porous carbon layer that conducts the electronic current and allows gas diffusion to the catalyst/membrane interface and liquid water to be extracted from the catalyst layer, and the oxygen reduction reaction is the rate-determining step. Noble metals such as Pt, Pd, and Au, are mainly catalysts for PEMFC, but these are very expensive, and Pt-based electatalysts and their associated catalyst layers contribute over 55% of the total cost of fuel cells. Thus, exploring new catalysts withl platinum or are platinum-free is in high demands [46, 47].

9.3.1 Pt-Based Catalysts

Pt-based catalysts are the mainly catalysts used in current fuel cells. It has a face-centered cubic (fcc) crystal structure with a lattice parameter of 3.93 Å, and it surface energies (γ) of the low-index crystallographic planes are in the order of $\gamma(111) < \gamma(100) < \gamma(110)$. Usually, nano-sized Pt-based catalysts have high specific surface area, and are high active than bulk, while the exposed facets, morphologies as well as composites affect the catalytic activities obviously [48]. Thus, synthesis of new nanostructured materials is a topic of intense research and development to optimize the existing Pt NP catalysts [49].

In 2007, Xia et al. at Washington University described the synthesis of platinum nanoparticles of an unusual shape (Fig. 9.4). Remarkably, the crystals were concave nanocubes enclosed by high index facets, including {510}, {720}, and/or {830} surfaces. Their as-prepared Pt concave nanocubes exhibited substantially enhanced specific activity compared with those of Pt nanocubes, cuboctahedra, and commercial Pt/C catalysts that are bounded by low index facets, such as {100} and {111} toward the oxygen reduction reaction, and have significant potential in fuel cell application [50].

Meanwhile, the Pt alloys have also been widely explored. Yang et al. synthesized Pt–Pd bimetallic heteronanostructures by using Pd as a metallic support for Pt nanoparticles (Fig. 9.5). They found that the alloys exhibited high reactivities and improved stabilities, which could be to the favorable interfacial structures between Pt and Pd supports, as well as the larger than usual overall particle size of Pt-on-Pd nanostructures, which prevented the small Pt from dissolution in the oxygen reduction reaction.

9.3.2 Nonnoble Metal Catalysts

Though Pt-based materials have long been investigated as active catalysts for oxygen reduction reaction, the large-scale application of fuel cells has been hampered by its high cost and the inadequacy of this metal. Recently, platinum-free catalysts for the oxygen reduction reaction have attracted enormous interest as an alternative to platinum-based catalysts. These mainly include three kinds of materials: organometallic complexes, nitrogen-doped carbon-supported metal ions, and transition metal oxides and chalcogenides

FIGURE 9.4   TEM (a) and HRTEM images (b,c), and SAED (d) of the Pt concave nanocubes recorded along [001] direction. *Source*: Reproduced with permission from Lim et al. [50].

[51, 52]. For example, Chen's group designed and synthesized a new highly durable iron phtalocyanine based nonprecious oxygen reduction reaction (ORR) catalyst (Fe–SPc) for polymer electrolyte membrane fuel cells (Fig. 9.6).

In 2010, Dai et al. synthesized nitrogen-doped graphene, and used it as a metal-free electrode in alkaline fuel cells, which showed a much better electrocatalytic activity, long-term operation stability, and tolerance to crossover effect than platinum for oxygen reduction (Fig. 9.7).

Beside organometallic complexes and nitrogen-doped carbon-supported metal ions, metal oxides or transition metal chalcogenides with high stability may be considered as potential alternatives to traditional Pt-based cathode catalysts.

Dai et al. in Stanford University reported a hybrid material consisting of Co_3O_4 nanocrystals grown on reduced grapheme oxide as a high performance bi-fuctional catalyst for the oxygen reduction reaction and oxygen evolution

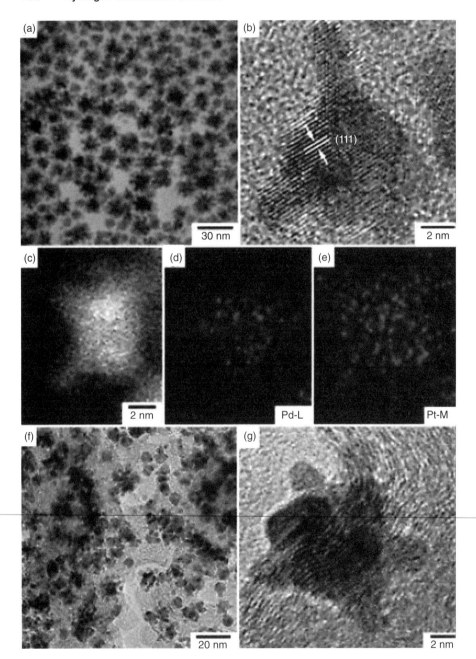

FIGURE 9.5   TEM (a), HRTEM (b), and HAADF-STEM (c) images, as well as elemental mappings of Pd (d) and Pt (e) metals in Pt–Pd bimetallic nanoparticles, TEM (f) and HRTEM (g) images of carbon-supported Pt–Pd bimetallic catalysts. *Source*: Reproduced with permission from Peng et al. [50].

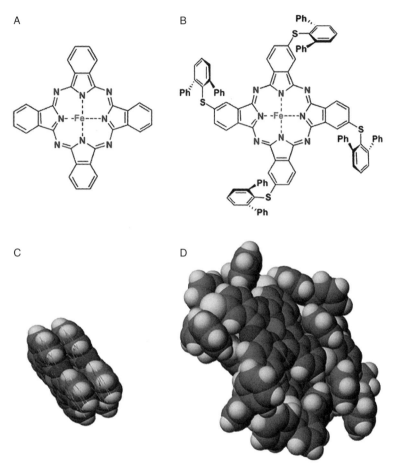

FIGURE 9.6 Atomic structure and the space filling stacking model of Fe–Pc (A,C) and Fe–SPc (B,D). *Source*: Reproduced with permission from Wu et al. [53]. (See color insert.)

reaction (Fig. 9.8). Though the CO_3O_4 and graphenen oxide alone has little catalytic activity, their hybrid exhibits an unexpected, surprisingly high oxygen reduction reaction performance.

Anderson et al. calculated the adsorption energies of the reactants, reaction intermediates, and products for water oxidation and O_2 reduction in acid on the (008), (002), and (202) surfaces of Co_9S_8 using the Vienna ab initio simulation program (VASP) (Fig. 9.9). They found that the partially OH-covered (202) surface is active toward O_2 reduction. On this surface, water cannot block the active sites, and heat loss from O_2 dissociative chemisorptions necessitates the overpotential for four-electron reaction.

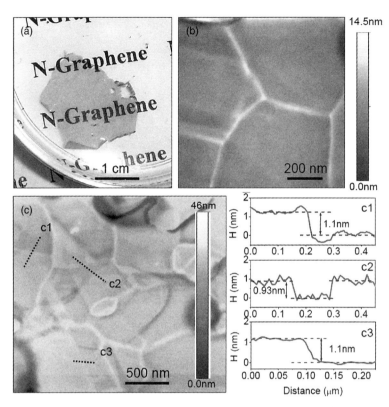

FIGURE 9.7 A digital photo image, AFM, and corresponding height analyses of the nitrogen doped grapheme. *Source*: Reproduced with permission from Gong et al. [54]. (See color insert.)

9.4 FUEL PROCESSING

With hydrogen as the main fuel for fuel cells, its reforming and storage are important issue [56]. There are a number possible sources for hydrogen generation, including fossil fuels, nuclear energy, wind energy, water electricity, and solar energy, as well as biomass that can be used to directly or indirectly produce hydrogen, as shown in Figure 9.10 [57]. Meanwhile, the primary techniques for converting fossil fuels into hydrogen are steam reforming (SR), partial oxidation (PO) reforming, and autothermal reforming (ATR), as discussed in Chapter 2. Other generation and storage methods are covered in other chapters of this book.

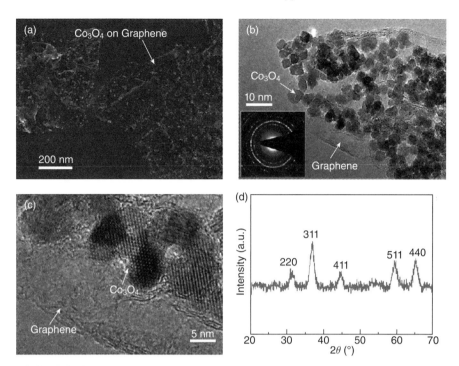

FIGURE 9.8   SEM (a), low (b) and high (c) magnification TEM images, as well as XPS spectrum (d) of Co_3O_4/grapheme hybrid materials. *Source*: Reproduced with permission from Liang et al. [55].

9.5 APPLICATIONS OF FUEL CELLS

The applications of fuel cells focus on propulsion of vehicles, stationary and portable power generation [26, 58, 59]. Among them, the vehicular application fuel cell systems are the major applications because of their potential huge impact on the environment, for example, the control of emission of the greenhouse gases. Most major motor companies work solely on PEMFCs due to their high power density and excellent dynamic characteristics as compared to other types of fuel cells [60]. Fuel-cell vehicles (FCVs) have been developed and demonstrated, for example, General Motors (GM) Hydrogen 1, Ford Demo IIa (Focus), DaimlerChrysler NeCar4a, Honda FCX-V3, Toyota FCHV, Nissan Xterra FCV, VW Bora HyMotion, and Hyundai Santa Fe FCV. Automakers such as Toyota, Honda, Hyudai, Daimler, and GM have announced plans of commercializing their FCVs by 2015 [61].

Among all applications for fuel cells, the transportation application involves the most stringent requirements regarding volumetric and gravimetric power density, reliability, and costs [62]. Since a widespread hydrogen

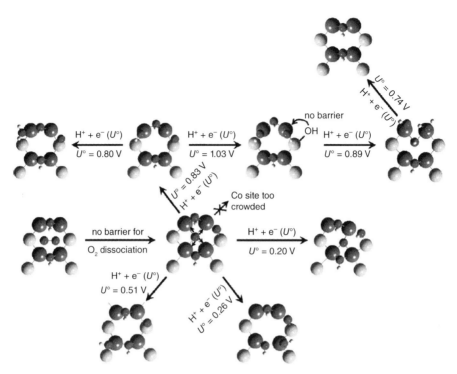

FIGURE 9.9   Top view of reduction pathways examined for the adsorbed O atoms on the partially OH(ads)-covered (202) surface of Co_9S_8. *Source*: Reproduced with permission from Sidik and Anderson [46]. (See color insert.)

retail infrastructure will not be available in the near future, car manufacturers consider a liquid fuel as the best option for a short-term market introduction of a fuel cell propulsion system [63]. Furthermore, the higher energy density of a liquid fuel guarantees a driving range similar to that of internal combustion engine vehicles [64]. The fuel favored by many car manufacturers is methanol from which hydrogen can be produced onboard by SR [56]. The reforming of the fuel, however, leads to slower response times, and extensive gas clean-up procedures need to be carried out to supply the fuel cell with high-grade hydrogen.

Some experts believe that fuel cell cars will never become economically competitive with other technologies or that it will take decades for them to become profitable [61, 65]. In July 2011, the Chairman and CEO of GM, Daniel Akerson, stated that while the cost of hydrogen fuel cell cars is decreasing: "The car is still too expensive and probably won't be practical until the 2020-plus period, I don't know" [56]. In 2013, Lux Research, Inc. issued a report that stated: "The dream of a hydrogen economy . . . is no nearer"[57]. It concluded that "capital cost . . . will limit adoption to a mere

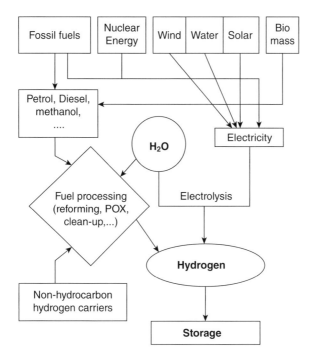

FIGURE 9.10   Possible production paths of hydrogen fuel. *Source*: Reproduced with permission from Carrette et al. [4].

5.9 GW" by 2030, providing "a nearly insurmountable barrier to adoption, except in niche applications."

9.6 SUMMARY

The fuel cell will change our lives as much as the computer has done and it may replacethe combustion engine one day. The future of power generation and transportation will change dramatically when fuel cells become the main producers of clean energy [62]. The car will no longer be an environmental burden, and the electrical grid will be subdivided to the neighborhood and building level. The quality and reliability of this alternative energy source will be superior to the current grid design. Fuel cells and hydrogen technology will potentially make all of this possible, and with little or no pollution.

Though the fuel cell has a bright future, there are still many problems with fuel cells. For example, hydrogen has some limitations that make it impractical for use in many applications. It is difficult to store and distribute, so it would be much more convenient if fuel cells could use fuels that are more readily available. However, as energy costs rise, the relative advantage of the high energy conversion efficiency of fuel cells will grow, and the business cases will strengthen accordingly (and this will occur alongside the expected reductions in manufacturing cost). In addition to being more efficient in end-use, development of fuel cells can also stimulate and ultimately lead to increased efficiency in energy conversion and harvesting.

REFERENCES

1. Pollet, B.G., Staffell, I., Shang, J.L. Current status of hybrid, battery and fuel cell electric vehicles: From electrochemistry to market prospects. *Electrochimica Acta* **2012**, *84*, 235–249.

2. Smil, V. Tomorrow's energy: Hydrogen, fuel cells, and the prospects for a cleaner planet. *Environment and Planning A* **2002**, *34*(12), 2260–2261.

3. Friedrich, K.A., Fuel Cells. *Bwk* **2012**, *64*(4), 110–116.

4. Carrette, L., Friedrich, K.A., Stimming, U. Fuel cells: fundamentals and applications. *Fuel Cells* **2001**, *1*(1), 5–39.

5. Carrette, L., Friedrich, K.A., Stimming, U. Fuel cells: Principles, types, fuels, and applications. *Chemphyschem* **2000**, *1*(4), 162–193.

6. Cao, D.X., Sun, Y., Wang, G.L. Direct carbon fuel cell: Fundamentals and recent developments. *Journal of Power Sources* **2007**, *167*(2), 250–257.

7. Wang, Y., Chen, K.S., Mishler, J., Cho, S.C., Adroher, X.C. A review of polymer electrolyte membrane fuel cells: Technology, applications, and needs on fundamental research. *Applied Energy* **2011**, *88*(4), 981–1007.

8. Sun, S., Jusys, Z., Behm, R.J. Electrooxidation of ethanol on Pt-based and Pd-based catalysts in alkaline electrolyte under fuel cell relevant reaction and transport conditions. *Journal of Power Sources* **2013**, *231*, 122–133.

9. Stoica, D., Ogier, L., Akrour, L., Alloin, F., Fauvarque, J.F. Anionic membrane based on polyepichlorhydrin matrix for alkaline fuel cell: Synthesis, physical and electrochemical properties. *Electrochimica Acta* **2007**, *53*(4), 1596–1603.

10. Lee, S.M., Kim, J.H., Lee, H.H., Lee, P.S., Lee, J.Y. The characterization of an alkaline fuel cell that uses hydrogen storage alloys. *Journal of the Electrochemical Society* **2002**, *149*(5), A603–A606.

11. Wang, J.H., Li, S.H., Zhang, S.B. Novel hydroxide-conducting polyelectrolyte composed of an poly(arylene ether sulfone) containing pendant quaternary guanidinium groups for alkaline fuel cell applications. *Macromolecules* **2010**, *43*(8), 3890–3896.

12. Adams, L.A., Poynton, S.D., Tamain, C., Slade, R.C.T., Varcoe, J.R. A carbon dioxide tolerant aqueous-electrolyte-free anion-exchange membrane alkaline fuel cell. *Chemsuschem* **2008**, *1*(1–2), 79–81.

13. Sanchez, D.G., Hiesgen, R., Wehl, I., Friedrich, K.A. Correlation of oscillation of polymer electrolyte membrane fuel cells at low cathode humidification with nanoscale membrane properties. *Batteries and Energy Technology (General): 219th Ecs Meeting* **2011**, *35*(32), 41–54.

14. Gulzow, E., Schulze, M., Friedrich, K.A., Fischer, P., Bettermann, H. Local in-Situ Analysis of Pem Fuel Cells by Impedance Spectroscopy and Raman Measurements. *Fuel Cell Seminar 2010* **2011**, *30*(1), 65–76.

15. Sanchez, D.G., Diaz, D.G., Hiesgen, R., Wehl, I., Friedrich, K.A. Oscillations of PEM fuel cells at low cathode humidification. *Journal of Electroanalytical Chemistry* **2010**, *649*(1–2), 219–231.

16. Zhang, S.S., Yuan, X.Z., Hin, J.N.C., Wang, H.J., Friedrich, K.A., Schulze, M. A review of platinum-based catalyst layer degradation in proton exchange membrane fuel cells. *Journal of Power Sources* **2009**, *194*(2), 588–600.

17. Silveira, J.L., Leal, E.M., Ragonha, L.F. Analysis of a molten carbonate fuel cell: cogeneration to produce electricity and cold water. *Energy* **2001**, *26*(10), 891–904.

18. Schulze, M., Gulzow, E., Friedrich, K.A. Low pressure test facility for polymer electrolyte membrane fuel cells and first measurements. *Fuel Cell Seminar 2007* **2008**, *12*(1), 187–197.

19. Costamagna, P., Srinivasan, S. Quantum jumps in the PEMFC science and technology from the 1960s to the year 2000: Part II. Engineering, technology development and application aspects. *Journal of Power Sources* **2001**, *102*(1–2), 253–269.

20. Oezaslan, M., Hasche, F., Strasser, P. Oxygen electroreduction on PtxCo1-x and PtxCu1-x alloy nanoparticles for basic and acidic pem fuel cell. *Polymer Electrolyte Fuel Cells 11* **2011**, *41*(1), 1659–1668.

21. Hasche, F., Fellinger, T.P., Oezaslan, M., Paraknowitsch, J.P., Antonietti, M., Strasser, P. Mesoporous nitrogen doped carbon supported platinum PEM fuel cell electrocatalyst made from ionic liquids. *Chemcatchem* **2012**, *4*(4), 479–483.

22. Hasche, F., Oezaslan, M., Strasser, P. Activity, Stability, and degradation mechanisms of dealloyed PtCu3 and PtCo3 nanoparticle fuel cell catalysts. *Chemcatchem* **2011**, *3*(11), 1805–1813.

23. Strasser, P. Fuel cell catalyst particles have platinum-rich shell, copper core. *Advanced Materials & Processes* **2008**, *166*(1), 13–13.

24. Strasser, P. Dealloyed core-shell fuel cell electrocatalysts. *Reviews in Chemical Engineering* **2009**, *25*(4), 255–295.

25. Wang, Y.J., Wilkinson, D.P., Zhang, J.J. Noncarbon support materials for polymer electrolyte membrane fuel cell electrocatalysts. *Chemical Reviews* **2011**, *111*(12), 7625–7651.

26. Debe, M.K. Electrocatalyst approaches and challenges for automotive fuel cells. *Nature* **2012**, *486*(7401), 43–51.

27. Mani, P., Srivastava, R., Strasser, P. Dealloyed Pt-Cu core-shell nanoparticle electrocatalysts for use in PEM fuel cell cathodes. *Journal of Physical Chemistry C* **2008**, *112*(7), 2770–2778.

28. Srivastava, R., Mani, P., Hahn, N., Strasser, P. Efficient oxygen reduction fuel cell electrocatalysis on voltammetrically dealloyed Pt-Cu-Co nanoparticles. *Angewandte Chemie-International Edition* **2007**, *46*(47), 8988–8991.

29. Stonehart, P. Development of advanced noble metal-alloy electrocatalysts for phosphoric-acid fuel-cells (PAFC). *Berichte Der Bunsen-Gesellschaft-Physical Chemistry Chemical Physics* **1990**, *94*(9), 913–921.

30. Ganguly, S., Das, S., Kargupta, K., Bannerjee, D. Optimization of performance of phosphoric acid fuel cell (PAFC) Stack using reduced order model with integrated space marching and electrolyte concentration inferencing. *11th International Symposium on Process Systems Engineering, Pts A and B* **2012**, *31*, 1010–1014.

31. Watanabe, M., Tsurumi, K., Mizukami, T., Nakamura, T., Stonehart, P. Activity and stability of ordered and disordered Co-Pt alloys for phosphoric-acid fuel-cells. *Journal of the Electrochemical Society* **1994**, *141*(10), 2659–2668.

32. Zhang, H.C., Lin, G.X., Chen, J.C., Multi-objective optimisation analysis and load matching of a phosphoric acid fuel cell system. *International Journal of Hydrogen Energy* **2012**, *37*(4), 3438–3446.

33. Zhang, H.S., Wang, L.J., Weng, S.L., Su, M. Control performance study on the molten carbonate fuel cell hybrid systems. *Proceedings of the Asme Turbo Expo* **2008**, *2*, 611–618.

34. Antolini, E. The stability of molten carbonate fuel cell electrodes: A review of recent improvements. *Applied Energy* **2011**, *88*(12), 4274–4293.

35. Bychin, V.P., Zvezdkin, M.A., Samatov, O. M. Porous nickel anode for molten-carbonate electrolyte fuel-cell. *Russian Journal of Electrochemistry* **1993**, *29*(11), 1173–1176.

36. Zhao, M.S., Sun, C.Y. A novel anode material for molten carbonate fuel cell. *Proceedings of the 6th International Symposium on Molten Salt Chemistry and Technology* **2001**, 402–408.

37. Batra, V.S., Maudgal, S., Bali, S., Tewari, P.K. Development of alpha lithium aluminate matrix for molten carbonate fuel cell. *Journal of Power Sources* **2002**, *112*(1), 322–325.

38. Baumgartner, C.E. NiO Solubility in molten Li/K carbonate under molten carbonate fuel-cell cathode environments. *Journal of the Electrochemical Society* **1984**, *131*(8), 1850–1851.

39. Travis, R.P., Hart, J., Costamagna, P., Agnew, G.D., Attia, O. Fuel cell system reliability for a pressurised integrated-plannar SOFC. *Proceedings of the 3rd International Conference on Fuel Cell Science, Engineering, and Technology* **2005**, 335–340.

40. Gemmen, R., Abernathy, H., Gerdes, K., Koslowske, M., McPhee, W.A., Tao, T. Fundamentals of liquid tin anode solid oxide fuel cell (Lta-Sofc) operation. *Advances in Solid Oxide Fuel Cells V* **2010**, *30*(4), 37–52.

41. Costamagna, P., Costa, P., Antonucci, V. Micro-modelling of solid oxide fuel cell electrodes. *Electrochimica Acta* **1998**, *43*(3–4), 375–394.

42. Magistri, L., Trasino, F., Costamagna, P. Transient analysis of solid oxide fuel cell hybrids - Part I: Fuel cell models. *Journal of Engineering for Gas Turbines and Power-Transactions of the Asme* **2006**, *128*(2), 288–293.

43. Costamagna, P., Magistri, L., Massardo, A.F. Design and part-load performance of a hybrid system based on a solid oxide fuel cell reactor and a micro gas turbine. *Journal of Power Sources* **2001**, *96*(2), 352–368.

44. Costamagna, P., Selimovic, A., Del Borghi, M., Agnew, G. Electrochemical model of the integrated planar solid oxide fuel cell (IP-SOFC). *Chemical Engineering Journal* **2004**, *102*(1), 61–69.

45. Mabena, L.F., Modibedi, R.M., Ray, S.S., Coville, N.J. Ruthenium supported on nitrogen-doped carbon nanotubes for the oxygen reduction reaction in alkaline media. *Fuel Cells* **2012**, *12*(5), 862–868.

46. Sidik, R.A., Anderson, A.B. Co_9S_8 as a catalyst for electroreduction of O_2: Quantum chemistry predictions. *Journal of Physical Chemistry B* **2006**, *110*(2), 936–941.

47. Morozan, A., Jousselme, B., Palacin, S. Low-platinum and platinum-free catalysts for the oxygen reduction reaction at fuel cell cathodes. *Energy & Environmental Science* **2011**, *4*(4), 1238–1254.

48. Zhang, J., Sasaki, K., Sutter, E., Adzic, R.R. Stabilization of platinum oxygen-reduction electrocatalysts using gold clusters. *Science* **2007**, *315*(5809), 220–222.

49. Bing, Y.H., Liu, H.S., Zhang, L., Ghosh, D., Zhang, J.J. Nanostructured Pt-alloy electro-catalysts for PEM fuel cell oxygen reduction reaction. *Chemical Society Reviews* **2010**, *39*(6), 2184–2202.

50. Peng, Z.M., Yang, H., Synthesis and oxygen reduction electrocatalytic property of Pt-on-Pd bimetallic heteronanostructures. *Journal of American Chemical Society* **2009**, *131*, 7542–7543.

51. Collman, J.P., Devaraj, N.K., Decreau, R.A., Yang, Y., Yan, Y.L., Ebina, W., Eberspacher, T.A., Chidsey, C.E.D. A cytochrome c oxidase model catalyzes oxygen to water reduction under rate-limiting electron flux. *Science* **2007**, *315*(5818), 1565–1568.

52. Gao, M.R., Jiang, J., Yu, S.H. Solution-based synthesis and design of late transition metal chalcogenide materials for oxygen reduction reaction (ORR). *Small* **2012**, *8*(1), 13–27.

53. Wu, G., More, K.L., Johnston, C.M., Zelenay, P. High-Performance electrocatalysts for oxygen reduction derived from polyaniline, iron, and cobalt. *Science* **2011**, *332*(6028), 443–447.

54. Gong, K.P., Du, F., Xia, Z.H., Durstock, M., Dai, L.M. Nitrogen-doped carbon nanotube arrays with high electrocatalytic activity for oxygen reduction. *Science* **2009**, *323*(5915), 760–764.

55. Liang, Y.Y., Li, Y.G., Wang, H.L., Zhou, J.G., Wang, J., Regier, T., Dai, H.J. Co_3O_4 nano-crystals on graphene as a synergistic catalyst for oxygen reduction reaction. *Nature Materials* **2011**, *10*(10), 780–786.

56. Schlapbach, L., Zuttel, A. Hydrogen-storage materials for mobile applications. *Nature* **2001**, *414*(6861), 353–358.

57. Ahmed, S., Krumpelt, M. Hydrogen from hydrocarbon fuels for fuel cells. *International Journal of Hydrogen Energy* **2001**, *26*(4), 291–301.

58. Friedrich, K.A., Kallo, J., Schirmer, J. Fuel cells for aircraft application. *ICNMM 2009, Pts A–B* **2009**, 1231–1240.

59. Dyer, C.K. Fuel cells for portable applications. *Journal of Power Sources* **2002**, *106*(1–2), 31–34.

60. Ahn, S.Y., Eom, S.Y., Rhie, Y.H., Sung, Y.M., Moon, C.E., Choi, G.M., Kim, D.J. Application of refuse fuels in a direct carbon fuel cell system. *Energy* **2013**, *51*, 447–456.

61. Finsterwolder, F. Where goes the drive? Fuel cell-component for the automobile. *Bwk* **2008**, *60*(1–2), S8–S9.

62. Harvey, D. The fuel cell for the automobile: Some deflating truths. *Tce* **2004**, (760), 34–36.

63. Heinzel, A., Hebling, C., Zedda, M. Fuel cells with restricted performances for portable applications. *Stationary Fuel Cell Power Plants: Commercialization* **2001**, *1596*, 97–106.

64. Trimm, D.L., Onsan, Z.I. Onboard fuel conversion for hydrogen-fuel-cell-driven vehicles. *Catalysis Reviews: Science and Engineering* **2001**, *43*(1–2), 31–84.

65. Li, Z.F. Research on energy management strategy of fuel cell/battery hybrid electric automobile. *ICMET 09: Proceedings of the 2009 International Conference on Mechanical and Electrical Technology* **2009**, 61–64.

10

Hydrogen Utilization in Chemical Processes

10.1 BACKGROUND

We have reviewed hydrogen utilization for direct combustion and fuel cells in the previous chapters. In fact, hydrogen gas has also been extensively used in different industrial chemical processes. For example, large amounts of hydrogen gas are required for the processing (or upgrading) of fossil fuel, including hydrocracking and hydroprocessing. Hydrogen is also used for chemical manufacturing processes, such as Haber process to synthesize ammonia and other nitrogen-based fertilizers, such as ammonium nitrate. Superconductor and semiconductor industries also use hydrogen. In this chapter, we will highlight some of the major applications of hydrogen in these chemical processes.

10.2 HYDROGEN UTILIZATION IN PETROLEUM INDUSTRY

10.2.1 Hydrocracking

Hydrocracking is an important process to produce fossil fuel from crude oil [1]. The crude oil, composed of high boiling point and high molecular weight hydrocarbons are cracked by hydrogen gas to produce a number of low

Hydrogen Generation, Storage, and Utilization, First Edition. Jin Zhong Zhang, Jinghong Li, Yat Li, and Yiping Zhao.
© 2014 John Wiley & Sons, Inc. Published 2014 by John Wiley & Sons, Inc.

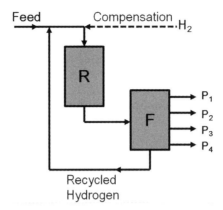

FIGURE 10.1 Schematic diagram of a typical of single stage of hydrocracking system. R, F, and P are reactor, fractionators, and product, respectively. *Source*: Reproduced with permission from Ward [2].

boiling point and low molecular weight hydrocarbons, such as gasoline and diesel oil, which can be used by the general public. A schematic diagram of a hydrocracking system is showed in Figure 10.1. Different hydrocarbons can be separated in a fractionator (F). The hydrogen gas used for hydrocracking can be recycled.

Noble metals such as palladium and platinum, and base metals such as molybdenum, tungsten, cobalt, and nickel are common catalysts used in hydrocracking. While the hydrogen gas is the essential component for the hydrocracking process, its reaction mechanism depends sensitively on the nature of the catalyst. The following provides several specific examples.

The first example is bifunctional catalysts that comprise a hydrogenation/dehydrogenation component, such as a noble metal, and a Brønsted acid component. Figure 10.2 illustrates a classical mechanism for the hydrocracking of an *n*-alkane. The reactant is first dehydrogenated on the metal sites to a mixture of *n*-alkenes, $n\text{-}C_iH_{2i}$. Then, they desorb from the metal sites and diffuse to Brønsted acid sites where they are protonated to the secondary alkylcarbenium ions, $n\text{-}C_iH_{2i+1}^+$. Carbenium ions undergo β-scission, and form smaller molecules and carbenium ions. These smaller species then diffuse to metal sites again and are hydrogenated to products.

The second example is monofunctional catalysts. Hydrogen gas and hydrocarbons are chemically adsorbed on the hydrogen absorption sites and reactant absorption sites of the same metal surface (Fig. 10.3). The hydrocracking reaction is initiated by carbon–carbon bond cleavage, followed by hydrogenation of the hydrocarbon fragments. This mechanism is also known as hydrogenolysis.

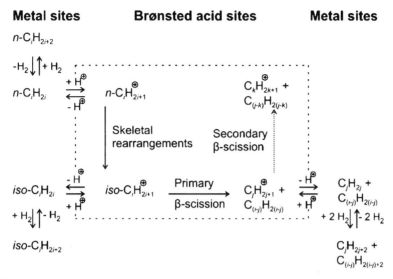

FIGURE 10.2   Classical mechanism of isomerization and hydrocracking of an alkane on a bifunctional catalyst comprising metal sites for dehydrogenation/hydrogenation and cracking sites. *Source*: Adapted with permission from Weitkamp [3].

$$2Z_H + H_2 \underset{k_{01}}{\overset{k_1}{\rightleftharpoons}} 2Z_H \bullet H$$

$$C_6H_5Cl + Z_R \xrightarrow{k_2} Z_R \bullet C_6H_5Cl$$

$$Z_R \bullet C_6H_5Cl + Z_R \xrightarrow{k_3} Z_R \bullet C_6H_5 + Z_R \bullet Cl$$

$$Z_R \bullet C_6H_5 + Z_H \bullet H \xrightarrow{k_4} Z_R + C_6H_6 + Z_H$$

$$Z_R \bullet Cl + Z_H \bullet H \xrightarrow{k_5} HCl + Z_R + Z_H$$

Overall Reaction: $C_6H_5Cl + H_2 \rightarrow C_6H_6 + HCl$

FIGURE 10.3   A proposed mechanism of hydrogenolysis of chlorobenzene on nickel-chromium metal. Z_H and Z_R indicate hydrogen absorption sites and reactant absorption sites, respectively [4].

The third example is Haag–Dessau hydrocracking, which is a nonclassical monomolecular cracking via carbonium ions [3]. This process usually happens on monofunctional acidic catalysts (such as acidic chabazite [5] and zeolite [6]). Figure 10.4 illustrates the mechanism of Haag–Dessau hydro-cracking. The initial reaction involves the protonation of alkane molecules (RH) to form carbonium ions (RH_2^+) on the surface of acidic catalyst. Car-bonium ions then collapse to give short alkanes (R_1H, usually methane and ethane) and carbenium ions (R_2^+). Carbenium ions then give back protons and desorb from catalyst to form alkenes, followed by hydrogenated by hydrogen.

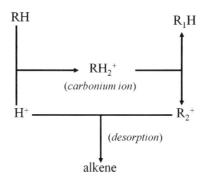

FIGURE 10.4   Hagg–Dessau cracking mechanism for an alkane molecule proceeding via a carbonium ion transition state [7].

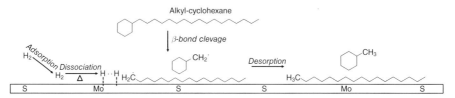

FIGURE 10.5   Proposed mechanism for hydrolysis of alkyl cyclohexanes on MoS_2 catalyst surface [8].

The fourth example is thermal hydrocracking (or hydropyrolysis). Even in the absence of a catalyst, thermal hydrocracking can be achieved at elevated temperatures (e.g., 500–600°C) and hydrogen pressure. As shown in Figure 10.5, the gaseous H_2 is first adsorbed on the surface of MoS_2 and dissociates into free radical H• when migrating to the Lewis acid sites. Simultaneously, the alkyl cyclohexanes are produced from β-scission of side chains under increasing thermal stress, probably with the assistance of catalysis of MoS_2. When the cleaved radical fragments encounters H•, termination reactions immediately take place to generate smaller molecules involving alkanes and cyclohexanes, which would be ultimately desorbed and expelled from the surface of catalyst [8].

10.2.2 Hydroprocessing

Petroleum products and natural gas typically contain a large amount of sulfur and nitrogen. The presence of sulfur and nitrogen impurities not only lowers the quality of the fuels but also causes severe air pollution by forming NO_x and SO_x gases during their combustion. These oxide gases are the major sources of acid rain. Hydroprocessing is the method designed to remove

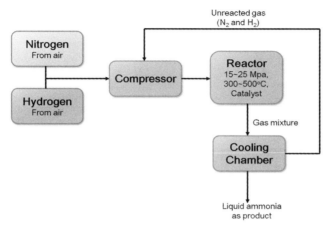

FIGURE 10.6 A flow chart illustration of the main components in a typical Haber process [11].

pollutants such as sulfur and nitrogen from chemical fuels. During this process, hydrogen gas will be used to hydrogenate sulfur and nitrogen compounds, and finally remove them in the form of H_2S and NH_3 gases [9].

10.3 HYDROGEN UTILIZATION IN CHEMICAL INDUSTRY

10.3.1 Ammonia Production: The Haber Process

The Haber process, also known as Haber–Bosch process, is the well-established method for production of ammonia using nitrogen and hydrogen as precursor gases. It was first reported by two German scientists Fritz Haber and Carl Bosch in the early twentieth century. This process has been refined and optimized for industrial-level production of ammonia and other nitrogen fertilizers. This is one of the most important processes in chemical industry, and significantly, it consumes almost 50% of the total global hydrogen production [9].

Figure 10.6 shows a schematic illustration of the Haber process. The Haber process takes nitrogen gas from air and combines it with hydrogen from natural gas to form ammonia:

$$N_2 + 3H_2 \underset{15-25\,\text{MPa},300\sim500^\circ\text{C}}{\overset{\text{Catalyst}}{\rightleftharpoons}} 2NH_3;\ \Delta H = -92.22\ \text{kJ}\cdot\text{mol}^{-1}. \quad (10.1)$$

Reaction pressure and temperature are typically set to be 15~25 MPa and 300~500°C. An optimal conversion efficiency of ~15% can be achieved in this single-step reaction. Unreacted gases will be separated from liquid ammonia and recycled. This synthetic loop can eventually achieve a remark-able conversion efficiency of 97% [10].

10.3.2   Hydrogenation of Unsaturated Hydrocarbons

Hydrogenation has been widely used for converting unsaturated organic compounds to saturated organic compounds in the presence of transition metal catalysts [9]. For instance, hydrogenation can convert alkene to alkane, alkyne to alkene or alkane, aldehyde to primary alcohol, ketone to secondary alcohol and nitrile to primary amine, and so on, as shown in Equation (10.2), Equation (10.3), Equation (10.4), and Equation (10.5) Lindlar reduction [12]. Therefore, hydrogenation is one of the most important reactions for organic synthesis.

$$H_2PdCl_4 + H_2O + CaCO_3 \rightarrow PdO/CaCO_3 \qquad (10.2)$$

$$PdO/CaCO_3 + HCOOH \rightarrow Pd/CaCO_3 \qquad (10.3)$$

$$Pd/CaCO_3 \xrightarrow{\text{Pb(OCOCH}_3)_2} \text{conditioned } Pd/CaCO_3 \qquad (10.4)$$

$$\qquad (10.5)$$

Moreover, hydrogen has also been used extensively to decrease the degree of unsaturation in vegetable oils. The reduction of unsaturation in vegetable oils can increase their melting points and enhance resistance to oxidation that enables preservation for a longer period of time. The degree of hydro-genation can be well controlled by varying the amount of hydrogen, reaction temperature and time, and the catalyst used (Fig. 10.7). For example, vege-table oil can be converted to margarine via partial hydrogenation. Some trans fats may be generated in the process of partial hydrogenation [13].

10.4   HYDROGEN UTILIZATION IN METALLURGICAL INDUSTRY

10.4.1   Ore Reduction

Hydrogen is a strong reducing agent that is used for the reduction of metallic ores to extract metal. Metals rarely exist in pure form. They are usually bonded to oxygen, sulfur, and occasionally to halogenides. Therefore, metals can be extracted from those metal ores through reduction process. Carbon,

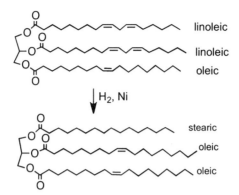

FIGURE 10.7 A schematic diagram illustrates partial hydrogenation of a triacylglycerol. The reactant is a typical kind of vegetable oil and the product is a typical component of margarine [14].

carbon monoxide, and hydrogen are commonly used as reduction agents. Among them, hydrogen has been avoided to be used for the production of metals such as iron, copper, and aluminum due to safety and cost considerations. Nevertheless, hydrogen has been commonly used for the extraction of two refractory metals, tungsten and molybdenum, because very pure metal powders can be obtained by hydrogen reduction of their native oxides [15].

For example, production of tungsten metal has almost been carried out exclusively by hydrogen reduction [16]. The reduction mechanism is depending on the moisture level in the environment. At low content of moisture, the reduction proceeds by solid-state diffusion of the oxygen out of the oxide, and can be represented by the following equation [15]:

$$WO_3(s) + 3H_2(g) \rightarrow W(s) + 3H_2O(g). \tag{10.6}$$

Nevertheless, another reaction path will be more favorable under higher temperature and high content of moisture. First, the oxygen-rich tungsten compounds will form $W_{20}O_{58}$ and $W_{18}O_{47}$ and eventually converted to WO_2. The WO_2 will further react with water to form $WO_2(OH)_2$, which can consequently react with hydrogen to produce pure metal tungsten:

$$WO_2(s) + H_2O(g) \rightarrow WO_2(OH)_2(g) + H_2(g) \tag{10.7}$$

$$WO_2(OH)_2(g) + 3H_2(g) \rightarrow W(s) + 4H_2O(g). \tag{10.8}$$

In practice, the grain size of tungsten powder can be controlled empirically by setting the temperature, oxide quantity, heating time, and H_2 flow rate. The range of size usually ranges from 0.1 to 60 μm [15].

FIGURE 10.8 SEM images show different morphologies of Mo metals obtained from hydrogen reduction under (left) low and (right) high moisture content. *Source*: Reproduced with permission from Schulmeyer [17].

As another example, molybdenum metal powder is produced industrially by reducing high purity molybdenum compounds such as molybdenum trioxide (MoO_3, grey-green powder), ammonium hexamolybdate ($(NH_4)_2Mo_6O_{19}$, yellow powder) and ammonium dimolybdate ($(NH_4)_2Mo_2O_7$, white powder), with hydrogen gas [15]. The mechanism of reduction consists of two stages [17]. The first stage involves a reaction from MoO_3 to MoO_2 via chemical vapor transport process. In the second stage, MoO_2 is further reduced to metal Mo:

$$MoO_2 + 2H_2 \rightarrow Mo + 2H_2O. \tag{10.9}$$

Different grain sizes and shapes of metal molybdenum can be obtained by adjusting moisture content of H_2 at entry (Fig. 10.8).

10.5 HYDROGEN UTILIZATION IN MANUFACTURING PROCESSES

10.5.1 Welding Gas: Oxy-Hydrogen Welding

Hydrogen has been used for welding as early as the beginning of the last century [18]. Hydrogen reacts with oxygen in a flame to form water, and simultaneously produces enormous energy for the welding process:

$$2H_2 + O_2 \longrightarrow 2H_2O; \Delta H = -572 \text{ kJ} \cdot \text{mol}^{-1}. \tag{10.10}$$

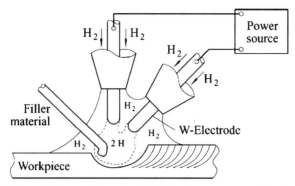

FIGURE 10.9   Schematic diagram illustrating the device for arc-atom welding. *Source*: Reproduced with permission from Suban et al. [18].

Before the introduction of acetylene, hydrogen was the basic combustible component in this kind of welding. Moreover, hydrogen gas has also been used for arc-atom welding (Fig. 10.9) [18].

As shown in Figure 10.9, an alternating arc is maintained between two tungsten electrodes, with hydrogen fed into the arc. When hydrogen is passed through the electric arc, the temperature in the arc core is sufficiently high to initiate dissociation of hydrogen gas to form atomic hydrogen:

$$H_2 \longleftrightarrow H + H; \Delta H = -422 \ kJ \cdot mol^{-1}. \qquad (10.11)$$

The energy required to break down the H–H covalent bond is provided by the arc. When the hydrogen atoms are recombined, they give the energy back and the flame at this point can reach as high as 3700°C and can thereby be used for welding. This process has been widely used for manual and automatic welding of metal sheets.

Recently, Tusek et al. has found that the static characteristic of the welding arc can be changed by adding different amounts of hydrogen in argon gas. The energy of the arc and the thermal and melting efficiency can be enhanced as a result of mixing hydrogen and other gases with different physical and chemical properties (e.g., thermoconductivity, enthalpy, and electric conductivity) [19].

10.5.2   Coolant

The thermal conductivity of hydrogen is substantially higher than that of other gases (Fig. 10.10). Moreover, it has high specific heat capacity, low

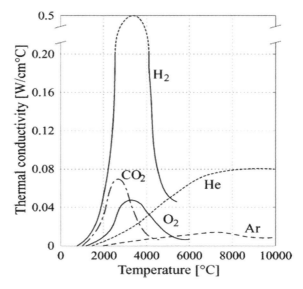

FIGURE 10.10   Thermal conductivity of various gases measured at different temperatures. *Source*: Reproduced with permission from Suban et al. [18].

density, and therefore low viscosity, which allows it to be used as high performance gaseous coolant. For example, hydrogen is commonly used in power stations as a coolant in turbo generators, which are designed to provide a low drag atmosphere and cooling for single-shaft and combined-cycle applications in combination with steam turbines [20]. In addition, there was attempt to use hydrogen as cooling gas in airplanes due to its low density. For example, Algarni et al. compared the performance of liquid hydrogen with methane as coolant for airplane [21]. The results suggested that liquid hydrogen is a superior candidate for coolant since it saves 10% of the initial total mass as compared with methane. Likewise, Ahmed et al. reported a comparative study for cooling an aerospace plane using liquid hydrogen, ammonia, and krypton [22]. They found that the required amount of liquid hydrogen is less than that of liquid NH_3 or Kr, and thus proved that liquid hydrogen is a better coolant. It should be noted that hydrogen is a highly flammable gas, and therefore any cooling system using hydrogen coolant should be completely sealed in order to guarantee only hydrogen gas is presented in the system to avoid explosion.

Liquid hydrogen is also a good coolant for superconductors, which will be discussed in Section 10.6.2.

10.6 HYDROGEN UTILIZATION IN PHYSICS

10.6.1 Lifting Gas

Hydrogen was once used as lifting gas for balloons and airships in the early 1900s. A famous airship filled with hydrogen was the Hindenburg airship, made in Germany in 1936. It was the longest class of flying machine and the largest airship by envelope volume. However, in 1937, the Hindenburg suddenly caught on fire and collapsed at Lakehurst Naval Air Station, New Jersey, while completing a single round trip passage to Rio de Janeiro. Although the exact cause of the Hindenburg disaster has not been determined yet, the drawback of using flammable hydrogen is clear. Thereafter, hydrogen was no longer used as filling gas for airships and it was replaced by helium gas.

10.6.2 Superconductor Industry

It is well known that superconductivity can be observed only at the temperature below a certain characteristic temperature, called critical temperature (T_c). Since most high temperature superconductors have T_c below 200 K [23], it is necessary to find a refrigerant to cool the superconductors. Liquid hydrogen has a boiling temperature of 20 K [24], which is even lower than that of liquid nitrogen (77 K) [25]. This makes liquid hydrogen a suitable candidate as a refrigerant for those superconductors with a T_c lower than the boiling point of liquid nitrogen, such as the magnesium diboride with a T_c of 39 K [24]. In addition, the current density of superconductors, for example, Ba–La–Cu–O ($T = 30$ K) [26] and Ag-sheathed Bi-2223 ($T = 20$ K) [27] superconductors, at liquid hydrogen temperature are higher than that at liquid nitrogen temperature.

10.6.3 Semiconductor Industry

Hydrogen can be easily introduced into semiconductors by occupying different sites in different charge states, forming complexes with impurities and defects [28, 29]. Hydrogen as an impurity can substantially affect the electronic properties of semiconductor materials. Significantly, hydrogen exhibits different behavior depending on the host crystal structure. As an isolated interstitial impurity, hydrogen can occupy a number of different lattice sites and modify the host structure, to the point of breaking host–atom bonds. It can act as either a donor (H^+) or an acceptor (H^-) [28]. Density-function theory calculation shows that hydrogen most stable in the positive charge state in p-type GaN, explaining the experimentally observed passivation of acceptors [30]. Likewise, H^- passivates donors in n-type GaN. In both cases, hydrogen always counteracts the prevailing conductivity and cannot act as a dopant in GaN [31].

There are some other cases in which hydrogen can also behave exclusively as donor or acceptor. For example, H^+ is the only stable charge state in ZnO, based on density-function theory calculation [32]. In this case, hydrogen will always give up its electron, thus increasing the concentration of free electrons and enhancing the electrical conductivity of ZnO. This is consistent with previously reported experimental results [28, 33, 34].

10.7 SUMMARY

In this chapter, we have briefly reviewed some examples of important applications of hydrogen gas beyond combustion and fuel cells. In the petroleum industry, hydrogen is mainly used for hydrocracking and hydroprocessing of crude oil, which produce large amounts of fossil fuels for commercial use. Hydrogen gas is also an essential material in the production of ammonia and nitrogen fertilizers through the Haber process, as well as hydrogenation of unsaturated hydrocarbons in chemical industry and organic synthesis. Moreover, hydrogen has been extensively used as reducing reagent for metal ore reduction. Lastly, hydrogen also can serve as a kind of lifting gas, a refrigerant for cooling superconductors and a dopant for semiconductors to modify their electronic properties. It is clear that hydrogen gas is essential for a number of extremely important chemical processes. Hydrogen sustainability is therefore critical for material supplies and economy growth. It is highly desired to develop environmentally friendly and sustainable methods for hydrogen production.

REFERENCES

1. Scherzer, J., Gruia, A.J. *Hydrocracking Science and Technology*, CRC Press, New York, 1996.

2. Ward, J.W. Hydrocracking processes and catalysts. *Fuel Process Technology*, **1993**, *35*(1–2), 55–85.

3. Weitkamp, J. Catalytic hydrocracking-mechanisms and versatility of the process. *ChemCatChem*, **2012**, *4*(3), 292–306.

4. Belokopytov, Y. Reaction mechanism in hydrogenolysis of chlorobenzene over nickel-chromium catalyst. *Theoretical and Experimental Chemistry*, **1998**, *34*(5), 280–282.

5. Bučko, T., Benco, L., Hafner, J., Ángyán, J.G. Monomolecular cracking of propane over acidic chabazite: An ab initio molecular dynamics and transition path sampling study. *Journal of Catalysis*, **2011**, *279*(1), 220–228.

6. (a) Babitz, S.M., Williams, B.A., Miller, J.T., Snurr, R.Q., Haag, W.O., Kung, H.H. Monomolecular cracking of n-hexane on Y, MOR, and ZSM-5 zeolites. *Applied Catalysis A: General*, **1999**, *179*(1–2), 71–86; (b) Chang, F.X., Wei, Y.X., Liu, X.B., Zhao, Y.F., Xu, L., Sun, Y., Zhang, D.Z., He, Y.L., Liu, Z.M. A mechanistic investigation of the coupled reaction of n-hexane and methanol over HZSM-5. *Applied Catalysis A: General*, **2007**, *328*(2), 163–173.

7. Kotrel, S., Knozinger, H., Gates, B.C. The Haag-Dessau mechanism of protolytic cracking of alkanes. *Microporous and Mesoporous Materials*, **2000**, *35–36*, 11–20.

8. He, K., Zhang, S., Mi, J., Chen, J., Cheng, L. Mechanism of catalytic hydropyrolysis of sedimentary organic matter with MoS_2. *Petroleum Science*, **2011**, *8*(2), 134–142.

9. Ramachandran, R., Menon, R.K. An overview of industrial uses of hydrogen. *International Journal of Hydrogen Energy*, **1998**, *23*(7), 593–598.

10. Appl, M., *Ammonia, 1. Introduction*, Wiley Publication, 2011.

11. Howard, F., James, H. *Chemical Reactor Design: For Process Plants*. In *Case Study 106: Ammonia Synthesis*, Austin, TX, John Wiley & Sons, 1977.

12. Lindlar, H., Dubuis, R. Palladium catalyst for partial reduction of acetylenes. *Organic Syntheses*, **1973**, *5*, 880.

13. Freeman, I.P. *Margarines and Shortenings*, Wiley Publication, 2000.

14. Breck, G., Bhatia, S. *Handbook of Industrial Oil and Fat Products*, CBS Publication, Delhi, 2008.

15. Luidold, S., Antrekowitsch, H. Hydrogen as a reducing agent: State-of-the-art science and technology. *Jom*, **2007**, *59*(6), 20–26.

16. Wiley-VCH, *Ullmann's Encyclopedia of Industrial Chemistry*, 6th ed., Wiley Publication, Weinheim, **2003**, Vol. 37.

17. Schulmeyer, W.V., Ortner, H.M. Mechanisms of the hydrogen reduction of molybdenum oxides. *International Journal of Refractory Metals & Hard Materials*, **2002**, *20*(4), 261–269.

18. Suban, M., Tusek, J., Uran, M. Use of hydrogen in welding engineering in former times and today. *Journal of Materials Processing Technology*, **2001**, *119*(1–3), 193–198.

19. Tusek, J., Suban, M. Experimental research of the effect of hydrogen in argon as a shielding gas in arc welding of high-alloy stainless steel. *International Journal of Hydrogen Energy*, **2000**, *25*(4), 369–376.

20. Nagano, S. Development of world's largest hydrogen-cooled turbine generator. *Power Engineering Society Summer Meeting, IEEE*, **2002**, *2*, 657–663.

21. Ahmed, Z., Ahmet, Z., Amro, M. Cooling Aerospace plane using hydrogen, ammonia and krypton. In *32nd Thermophysics Conference*, American Institute of Aeronautics and Astronaulics, Inc., 1997.

22. Algarni, A.Z., Sahin, A.Z., Yilbas, B.S., Ahmed, S.A. Cooling of Aerospace Plane Using Liquid-Hydrogen and Methane. *Journal of Aircraft*, **1995**, *32*(3), 539–546.

23. Ford, P.J. *The Rise of Superconductors*, CRC Press, Boca Raton, FL, 2005.

24. Hirabayashi, H., Makida, Y., Nomura, S., Shintomi, T. Feasibility of hydrogen cooled superconducting magnets. Applied Superconductivity, *IEEE Transactions*, **2006**, *16*(2), 1435–1438.

25. Yamashita, S., Hotate, K. Multiwavelength erbium-doped fibre laser using intracavity etalon and cooled by liquid nitrogen, *Electronics Letters*, **1996**, *32*(14), 1298–1299.

26. Bednorz, J.G., Muller, K.A. Possible highT c superconductivity in the Ba–La–Cu–O system, *Zeitschrift für Physik B-Condensed Matter*, **1986**, *64*(2), 189–193.

27. Kobayashi, S., Kaneko, T., Kato, T., Fujikami, J., Sato, K. A novel scaling of magnetic field dependencies of critical currents for Ag-sheathed Bi-2223 superconducting tape. *Physica C: Superconductivity*, **1996**, *258*(3–4), 336–340.

28. Van de Walle, C.G., Neugebauer, J. Universal alignment of hydrogen levels in semiconductors, insulators and solutions. *Nature*, **2003**, *423*(6940), 626–628.

29. Van de Walle, C.G., Neugebauer, J. Hydrogen in Semiconductors. *Annual Review of Materials Research*, **2006**, *36*(1), 179–198.

30. Neugebauer, J., Van de Walle, C. Hydrogen in GaN: Novel Aspects of a Common Impurity. *Physical Review Letters*, **1995**, *75*(24), 4452–4455.

31. Myers, S.M., Wright, A.F., Petersen, G.A., Seager, C.H., Wampler, W.R., Crawford, M.H., Han, J. Equilibrium state of hydrogen in gallium nitride: Theory and experiment. *Journal of Applied Physics*, **2000**, *88*(8), 4676–4687.

32. Van de Walle, C.G. Hydrogen as a cause of doping in zinc oxide. *Physical Review Letters*, **2000**, *85*, (5), 1012–1015.

33. Yang, P., Xiao, X., Li, Y., Ding, Y., Qiang, P., Tan, X., Mai, W., Lin, Z., Wu, W., Li, T., Jin, H., Liu, P., Zhou, J., Wong, C.P., Wang, Z.L. Hydrogenated ZnO Core-Shell Nanocables for Flexible Supercapacitors and Self-Powered Systems. *ACS Nano*, **2013**, *7*(3), 2617–26.

34. Hofmann, D.M., Hofstaetter, A., Leiter, F., Zhou, H.J., Henecker, F., Meyer, B.K., Orlinskii, S.B., Schmidt, J., Baranov, P.G. Hydrogen: A relevant shallow donor in zinc oxide. *Physical Review Letters*, **2002**, *88*(4), 045504.

Index

Hydrogen Generation, Storage, and Utilization, First Edition. Jin Zhong Zhang, Jinghong Li, Yat Li, and Yiping Zhao.
© 2014 John Wiley & Sons, Inc. Published 2014 by John Wiley & Sons, Inc.

191